Rethinking Maps

Maps are changing. They have become important and fashionable once more. *Rethinking Maps* brings together leading researchers to explore how maps are being rethought, made and used, and what these changes mean for working cartographers, applied mapping research, and cartographic scholarship. It offers a contemporary assessment of the diverse forms that mapping now takes and, drawing upon a number of theoretic perspectives and disciplines, provides an insightful commentary on new ontological and epistemological thinking with respect to cartography.

This book presents a diverse set of approaches to a wide range of map forms and activities in what is presently a rapidly changing field. It employs a multi-disciplinary approach to important contemporary mapping practices, with chapters written by leading theorists who have an international reputation for innovative thinking. Much of the new research around mapping is emerging as critical dialogue between practice and theory and this book has chapters focused on intersections with play, race and cinema. Other chapters discuss cartographic representation, sustainable mapping and visual geographies. It also considers how alternative models of map creation and use such as open-source mappings and map mashup are being creatively explored by programmers, artists and activists. There is also an examination of the work of various 'everyday mappers' in diverse social and cultural contexts.

This blend of conceptual chapters and theoretically directed case studies provides an excellent resource suited to a broad spectrum of researchers, advanced undergraduate and postgraduate students in human geography, GIScience and cartography, visual anthropology, media studies, graphic design and computer graphics. *Rethinking Maps* is a necessary and significant text for all those studying or having an interest in cartography.

Martin Dodge works at the University of Manchester as a Senior Lecturer in Human Geography researching the geography of cyberspace. He is the curator of a Web-based Atlas of Cyberspace (www.cybergeography.org/atlas) and has co-authored three books, *Mapping Cyberspace*, *Atlas of Cyberspace* and *Geographic Visualization*.

Rob Kitchin is Director of the National Institute of Regional and Spatial Analysis and Professor of Human Geography at the National University of Ireland, Maynooth. He has published twelve books and is the Managing Editor of *Social and Cultural Geography* and co-editor-in-chief of the *International Encyclopaedia of Human Geography*.

Chris Perkins is Senior Lecturer in Geography and Map Curator in the University of Manchester. His research interests focus on the social contexts of mapping and he is the author and editor of six books, including *World Mapping Today* and the *Companion Encyclopaedia of Geography*.

Routledge Studies in Human Geography

This series provides a forum for innovative, vibrant, and critical debate within Human Geography. Titles will reflect the wealth of research that is taking place in this diverse and ever-expanding field.

Contributions will be drawn from the main sub-disciplines and from innovative areas of work that have no particular sub-disciplinary allegiances.

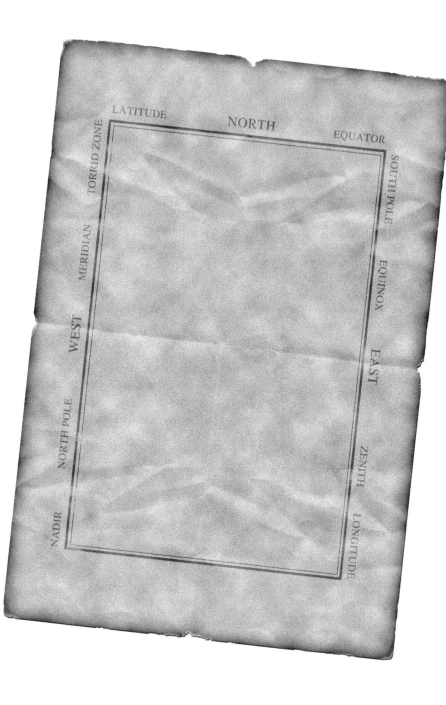

LATITUDE

NORTH

EQUATOR

TORRID ZONE

SOUTH POLE

MERIDIAN

EQUINOX

WEST

EAST

NORTH POLE

ZENITH

NADIR

LONGITUDE

Rethinking Maps

New frontiers in cartographic theory

Martin Dodge, Rob Kitchin and Chris Perkins

LONDON AND NEW YORK

First published 2009
Paperback edition first published 2011
by Routledge
2 Park Square, Milton Park, Abingdon, Oxon OX14 4RN

Simultaneously published in the USA and Canada
by Routledge
711 Third Avenue, New York, NY 10017

Routledge is an imprint of the Taylor & Francis Group, an informa business

British Library Cataloguing in Publication Data
A catalogue record for this book is available from the British Library

Library of Congress Cataloging in Publication Data
A catalog record for this book has been requested

ISBN: 978-0-415-46152-8 (hbk)
ISBN: 978-0-415-67667-0 (pbk)
ISBN: 978-0-203-87684-8 (ebk)

Typeset in Times New Roman by
Florence Production Ltd, Stoodleigh, Devon

Martin dedicates this book to his sisters
Alison and Susan

Contents

Illustrations

Figures

Tables

Contributors

Stuart C. Aitken
Department of Geography, San Diego State University San Diego, USA and Department of Geography and The Norwegian Centre for Child Research, Norwegian University of Science and Technology, Trondheim, Norway

Stuart Aitken is Professor of Geography at San Diego State University. His books include *Philosophies, People, Places and Practices* (with Gill Valentine; Sage 2004), *Geographies of Young People: The Morally Contested Spaces of Identity* (Routledge 2001), *Family Fantasies and Community Space* (Rutgers University Press 1998), *Place, Space, Situation and Spectacle: A Geography of Film* (with Leo Zonn; Rowman & Littlefield 1994) and *Putting Children in Their Place* (Association of American Geographers 1994). His interests include film and media, critical social theory, qualitative methods, public participation GIS, children, families and communities. Stuart is past co-editor of *The Professional Geographer* and current North American editor of *Children's Geographies*.

Tom Conley
Department of Romance Languages and Literatures, Harvard University, USA

Tom Conley is member of the Departments of Romance Languages and Visual/Environmental Studies at Harvard University. He studies cartography through literature, film and theory and is the author of *The Self-Made Map: Cartographic Writing in Early Modern France* (University of Minnesota Press 1996) and *Cartographic Cinema* (University of Minnesota Press 2007). He has contributed two essays to David Woodward (ed.) *The History of Cartography 3: The European Renaissance* (University of Chicago Press 2007) and is translator of Christian Jacob, *The Sovereign Map* (University of Chicago Press 2006).

Emmet Connolly
Dublin, Ireland

Emmet Connolly is an interaction designer and artist. His work encompasses open source programming, mapping methodologies, information visualization, citizen media and community activism. Emmet earned an MA in digital art and technology at the University of Plymouth and currently works for Google in Dublin.

Jim Craine
Department of Geography, California State University, Northridge, USA

Jim Craine is an Assistant Professor of Geography at California State University, Northridge. He specializes in the geography of media and also works applying geovisualization theory to digital and analogue cartography. He is co-editor of *Aether: The Journal of Media Geography* (www. aetherjournal.org).

Jeremy W. Crampton
Department of Geosciences, Georgia State University, USA

Jeremy Crampton is Associate Professor of Geography with research interests focused on the critical approaches to cartography and GIS, the politics of identity and new spatial media. His paper 'the biopolitical justification for geosurveillance' appeared in the *Geographical Review* in 2007, and a progress report on cartography and new spatial media was published in *Progress in Human Geography* in 2008. He is editor of *Cartographica* and also a section editor for cartography in the *International Encyclopedia of Human Geography* (Elsevier 2009) and author of the forthcoming book *Mapping: A Critical Introduction to Cartography and GIS* (Wiley-Blackwell).

Martin Dodge
School of Environment and Development, University of Manchester, UK

Martin Dodge is a Senior Lecturer in Human Geography and his research focuses primarily on the geography of cyberspace, particularly ways to map and visualize the Internet and the Web. He is the curator of a Web-based *Atlas of Cyberspace*s (www.cybergeography.org/atlas) and has co-authored two books, *Mapping Cyberspace* (Routledge 2000) and *Atlas of Cyberspace* (Addison-Wesley 2001) both with Rob Kitchin and recently co-edited a book on *Geographic Visualization* (Wiley 2008). He has a PhD from the University of London.

Georg Gartner
Institute of Geoinformation and Cartography, Vienna University of Technology, Austria

Georg Gartner is a Full Professor in the research group in cartography at the Vienna University of Technology. He holds graduate qualifications in geography and cartography from the University of Vienna and received his PhD and his Habilitation from the Vienna University of Technology. He was awarded a Fulbright grant to the University of Nebraska at Omaha in

1997 and a research visiting fellowship to the Royal Melbourne Institute of Technology in 2000 and to South China Normal University in 2006. He serves as Vice President of the International Cartographic Association and is the organizer of the International Symposia on Location Based Services & TeleCartography and co-editor of the book series *Lecture Notes on Geoinformation and Cartography* (Springer) and the *Journal of Location Based Services* (Taylor & Francis).

Michael F. Goodchild

National Center for Geographic Information and Analysis and Department of Geography, University of California, Santa Barbara, USA

Michael Goodchild is Professor of Geography at the University of California, Santa Barbara, and Director of spatial@ucsb. He received his BA degree in physics from Cambridge University in 1965 and his PhD in geography from McMaster University in 1969. He was Director of the National Center for Geographic Information and Analysis from 1991 to 1997. He was elected member of the National Academy of Sciences and Foreign Fellow of the Royal Society of Canada in 2002, and member of the American Academy of Arts and Sciences in 2006. His published books include *Accuracy of Spatial Databases* (Taylor & Francis 1989); *Geographical Information Systems: Principles and Applications* (Longman Scientific and Technical 1991); *Environmental Modeling with GIS* (Wiley 1996); *Scale in Remote Sensing and GIS* (CRC Press 1997); *Interoperating Geographic Information Systems* (Springer 1999); *Geographic Information Systems and Science* (Wiley 2001); *Uncertainty in Geographical Information* (CRC Press 2002); *Foundations of Geographic Information Science* (CRC Press 2003); S*patially Integrated Social Science* (OUP 2004); *GIS, Spatial Analysis and Modeling* (ESRI Press 2005); and *Geospatial Analysis: A Comprehensive Guide to Principles, Techniques and Software Tools* (Winchelsea Press 2007). His current research interests centre on geographic information science, spatial analysis, and uncertainty in geographic data.

Leila Harris

Department of Geography and Gaylord Nelson Institute for Environmental Studies, University of Wisconsin-Madison, USA

Leila Harris is Assistant Professor of Geography and Environmental Studies. She received her BA in Political Economy of Industralized Societies from the University of California-Berkeley and her MA and PhD in Geography from the University of Minnesota. Her publications to date include articles in journals such as *Environment and Planning D: Society and Space*, *World Development*, *Geoforum* and *Local Environment*. She works at the intersection of nature-society theory, gender studies and critical development studies. Among other research efforts, her work includes attention to critical cartography and socio-natures; social, political and institutional dimensions of environmental and developmental change in contemporary Turkey; emergent international and institutional trends with respect to water governance; and gender-environment debates.

Helen Hazen
Department of Geography, Macalester College, St Paul, USA

Helen Hazen is Visiting Professor of Geography at Macalester College. She is an environmental and medical geographer with field experience in the Americas, Africa and Oceania. After several years working in conservation in the tropics, she returned to academia to pursue interests related to national parks and international conversation agreements. Her research continues to focus on human–environment interactions, but has subsequently branched out to incorporate the geography of human health, particularly ecological approaches to health and disease, as a major interest.

Rob Kitchin
National Institute for Regional and Spatial Analysis and Department of Geography, National University of Ireland, Maynooth, Ireland

Rob Kitchin is Professor of Human Geography and Director of the National Institute of Regional and Spatial Analysis (NIRSA) at the National University of Ireland, Maynooth and Chair of the Management Board of the Irish Social Sciences Platform (ISSP). He has published sixteen books, is editor of the international journal *Social and Cultural Geography*, and co-editor-in-chief of the *International Encyclopedia of Human Geography* (Elsevier 2009).

John Krygier
Department of Geology and Geography, Ohio Wesleyan University, USA

John Krygier is Associate Professor of Geography. He teaches cartography, GIS and geography courses at Ohio Wesleyan University, and is the author, with Denis Wood, of *Making Maps: A Visual Guide to Map Design for GIS* (Guilford Press 2005). He blogs about cartographic arcane at http://makingmaps.net.

Chris Perkins
School of Environment and Development, University of Manchester, UK

Chris Perkins is Senior Lecturer in Geography. He is the author of four books including standard texts documenting the changing contexts of map availability (*World Mapping Today* with R.B. Parry; Bowker 1998) and has recently co-edited the second edition of the *Companion Encyclopaedia to Geography* (Routledge 2006). His research interests are centred on the different ways in which mapping may be employed and he is the first Chair of the International Cartographic Association's Commission on Maps in Society.

Amy D. Propen
Department of English and Humanities, York College of Pennsylvania, USA

Amy Propen is Assistant Professor of Rhetoric and Composition. Her areas of research include the intersections of rhetoric and geography, with a specific focus on visual and material rhetoric, space, and the body. She has published on critical GIS, visual rhetoric and mapping, and visual communication, and is currently preparing a book about understanding material rhetoric through critical approaches to mapping and locative technologies. She has a PhD in rhetoric and scientific and technical communication from the University of Minnesota and a BA degree in Geography and English from Clark University.

Dominica Williamson
Cornwall, UK

Dominica Williamson is a freelance designer and artist based in Cornwall. Her practice looks at how sustainable design solutions can be achieved. She was the design director for Green Map's *Green Map Atlas* in 2003–4 and joined the network in 2000. She was awarded a MSc in digital futures at the School of Computing, Communications and Electronics, University of Plymouth and obtained a BA first class honours in design studies from Goldsmiths College, University of London. Her latest work can be found at http://ecogeographer.com.

Denis Wood
Raleigh, North Carolina, USA

Denis Wood is an independent scholar and he was Professor of Design at North Carolina State University from 1974 to 1996. He curated The Power of Maps exhibitions for the Cooper-Hewitt National Museum of Design in New York and the Smithsonian in Washington. Author of *The Power of Maps* (Guilford Press 2002), *Home Rules* (with Robert Beck; Johns Hopkins University Press 1994), *Seeing Through Maps* (with Ward Kasier; ODT 2001), *Five Billion Years of Global Change* (Guilford Press 2004) and *Making Maps* (with John Krygier; Guilford Press 2005), his most recent book is *The Natures of Maps* (with John Fels; University of Chicago Press 2008). He has a PhD in geography from Clark University.

Preface

The Bellman's Speech

The Bellman himself they all praised to the skies –
Such a carriage, such ease and such grace!
Such solemnity, too! One could see he was wise,
The moment one looked in his face!
He had bought a large map representing the sea,
Without the least vestige of land:
And the crew were much pleased when they found it to be
A map they could all understand.
"What's the good of Mercator's North Poles and Equators,
Tropics, Zones, and Meridian Lines?"
So the Bellman would cry: and the crew would reply
"They are merely conventional signs!
"Other maps are such shapes, with their islands and capes!
But we've got our brave Captain to thank"
(So the crew would protest) "that he's bought us the best –
A perfect and absolute blank!"

(Lewis Carroll (1876), *The Hunting of the Snark*, London: Macmillan)

1   Thinking about maps

Rob Kitchin, Chris Perkins and
Martin Dodge[1]

Introduction

> A map is, in its primary conception, a conventionalized picture of the
> Earth's pattern as seen from above.
>
> (Raisz 1938)

> Every map is someone's way of getting you to look at the world his or
> her way.
>
> (Lucy Fellowes, Smithsonian curator, quoted in Henrikson 1994)

Given the long history of mapmaking and its scientific and scholarly traditions
one might expect the study of cartography and mapping theory to be relatively
moribund pursuits with long established and static ways of thinking about
and creating maps. This, however, could not be further from the truth. As
historians of cartography have amply demonstrated, cartographic theory and
praxis have varied enormously across time and space, and especially in recent
years. As conceptions and philosophies of space and scientific endeavour
have shifted so has how people come to know and map the world.

Philosophical thought concerning the nature of maps is of importance
because it dictates how we think about, produce and use maps; it shapes our
assumptions about how we can know and measure the world, how maps
work, their techniques, aesthetics, ethics, ideology, what they tell us about
the world, the work they do in the world, and our capacity as humans to
engage in mapping. Mapping is epistemological but also deeply ontological
– it is both a way of thinking about the world, offering a framework for
knowledge, and a set of assertions about the world itself. This philosophical
distinction between the nature of the knowledge claims that mapping is able
to make, and the status of the practice and artefact itself, is intellectually
fundamental to any thinking about mapping.

In this opening chapter we explore the philosophical terrain of contemporary
cartography, setting out some of the reasons as to why there is a diverse
constellation of map theories vying for attention and charting some significant
ways in which maps have been recently theorized. It is certainly the case

that maps are enjoying something of a renaissance in terms of their popularity, particularly given the various new means of production and distribution. New mapping technologies have gained the attention of industry, government and to some extent the general public keen to capitalize on the growing power, richness and flexibility of maps as organizational tools, modes of analysis and, above all, compelling visual images with rhetorical power. It is also the case that maps have become the centre of attention for a diverse range of scholars from across the humanities and social sciences interested in maps in and of themselves and how maps can ontologically and epistemologically inform other visual and representational modes of knowing and praxis. From a scientific perspective, a growing number of researchers in computer science and engineering are considering aspects of automation of design, algorithmic efficiency, visualization technology and human interaction in map production and consumption.

These initiatives have ensured that mapping theory over the past twenty years has enjoyed a productive period of philosophical and practical development and reflection. Rather than be exhaustive, our aim is to demonstrate the vitality of present thinking and practice, drawing widely from the literature and signposting relevant contributions among the chapters that follow. We start our discussion by first considering the dimensions across which philosophical differences are constituted. We then detail how maps have been theorized from within a representational approach, followed by an examination of the ontological and epistemological challenges of post-representational conceptions of mapping.

Dimensions across which map theory is constituted

A useful way of starting to understand how and why map theory varies is to explore some of the dimensions across which philosophical debate is made. Table 1.1 illustrates some important binary distinctions that strongly influence views on the epistemological and ontological status of mapping: judging a philosophy against these distinctions provides an often unspoken set of rules for knowing the world, or in our case, for arguing about the status of mapping. These distinctions are clearly related to each other. An emphasis upon the map as representation, for example, is also often strongly associated with the quest for general explanation, with a progressive search for order, with Cartesian distinctions between the map and the territory it claims to represent, with rationality, and indeed with the very act of setting up dualistic categories. By exploring how these dimensions work we can begin to rethink mapping and explain the complex variety of approaches described later in this book.

The mind–body distinction is often a fundamental influence on how people think about the world. If the mind is conceptualized as separate from the body then instrumental reason becomes possible: the map can be separated from the messy and subjective contingencies that flow from an embodied

Table 1.1 Rules for knowing the world: binary opposites around which ideas coalesce

Mind	Body	Structure	Agency
Empirical	Theoretical	Process	Form
Absolute	Relative	Production	Consumption
Nomothetic	Ideographic	Representation	Practice
Ideological	Material	Functional	Symbolic
Subjective	Objective	Immutable	Fluid
Essence	Immanence	Text	Context
Static	Becoming	Map	Territory

view of mapping. As such, science and reason become possible and a god-like view from nowhere can represent the world in an objective fashion, like a uniform topographic survey. On the other hand assuming a unity of mind and body and emphasizing the idea of embodied knowing focuses attention on different, more hybrid and subjective qualities of mapping, rendering problematic distinctions between the observer and observed.

The question of whether geographic knowledge is unique or whether the world might be subject to more general theorizing also has fundamental implications for mapping. An ideographic emphasis on uniqueness has frequently pervaded theorizing about mapping in the history of cartography: if each map were different, and described a unique place, searching for general principles that might govern design, or explain use would be doomed to fail. Instead, mapping becomes the ultimate expression of descriptive endeavour, an empirical technique for documenting difference. Artistic approaches to mapping that privilege the subjective may be strongly compatible with this kind of interpretation. On the other hand a more nomothetic approach, which emphasizes laws and denies idiosyncratic difference risks reifying artificially theorized models or generalizations while at the same time offering the possibility of scientific universalization. Many of the approaches described in the chapters by Goodchild and Gartner in this volume subscribe to this quest for order. Debate continues around the nature of map generalization and whether mapping is holistic or fragmentary, stochastic or regular, invariant or contingent, natural or cultural, objective or subjective, functional or symbolic, and so on. It is clear, however, that since World War II a number of different scientific orthodoxies have pervaded the world of Western academic cartographic research which almost all trade on the notion of searching for a common, universal approach. Yet, paradoxically, everyday ideas of geography and mapping as ideographic and empirical survive.

As we examine in detail later in the chapter, the idea of viewing maps as texts, discourses or practices emerged in the late 1980s, in stark opposition to the more practical and technologically driven search for generalization. These new theoretical ways of understanding mapping often emphasized the discursive power of the medium, stressing deconstruction, and the social and cultural work that cartography achieves. Here, the power of mapping becomes a more important consideration than the empirical search for verifiable generalization, and the chapters in this volume by Crampton, Harris and Hazen, and Propen consider some of these alternative approaches.

Structural explanations of the significance of mapping have also strongly influenced understandings of maps. Insights drawn might stem from class relations, from cultural practice, from psychoanalysis, or linguistics: for example, semiotic approaches to mapping have been a powerful and influential way of approaching the medium and its messages for academic researchers (see Krygier and Wood's chapter). There is an ongoing debate in relation to mapping over how the agency of an individual might be reconciled with this kind of approach, given that structural approaches often posit fundamental and inevitable forces underpinning all maps. There is also a continuing debate over the philosophical basis of the structural critique. For example, is it grounded in a materialist view of the world, or in a more ideological reading of the human condition.

The distinction between forces producing the world and the forces consuming it also has a strong resonance in philosophical debates around mapping. The cultural turn in academic geography encouraged a growing emphasis on the contexts in which maps operate, encouraging a shift away from theorizing about production and towards philosophies of mapping grounded in consumption. Here, the map reader becomes as important as the mapmaker. Technological change that reduced the significance of barriers to accessing data, and the democratization of cartographic practice have also encouraged this changed emphasis. Associated with this shift has been the increasingly nuanced drift towards poststructuralist ways of knowing the world, which distrust all-encompassing knowledge claims. Instead of a belief in absolute space, or a socially constructed world, an alternative way of understanding mapping has emphasized relativity and contingency in a universe where notions of reality come to be replaced by simulation and in which the play of images replaces visual work, or in which speed of change itself gains agency.

Representational cartography

Maps as truth

It is usually accepted that cartography as a scientific endeavour and industry seeks to represent as faithfully as possible the spatial arrangements of phenomena on the surface of the earth. The science of cartography aims to accurately capture relevant features and their spatial relations and to re-present a scaled abstraction of that through the medium of a map. Maps seek to be truth documents; they represent the world as it really is with a known degree of precision. Cartography as an academic and scientific pursuit then largely consists of theorizing how best to represent and communicate that truth (through new devices, e.g. choropleth maps, contour lines; through the use of colour; through ways that match how people may think, e.g. drawing on cognitive science).

This quest for producing truth documents has been the preoccupation for Western cartographers since the late Middle Ages, and especially with the need for accurate maps with respect to navigation, fighting wars and regulating property ownership. It was only in the 1950s, however, that the first sustained attempts began to emerge in the US to reposition and remould academic cartography as an entirely scientific pursuit. Up until then the history of cartography was a story of progress. Over time maps had become more and more precise, cartographic knowledge improved, and implicitly it was assumed that everything could be known and mapped within a Cartesian framework. The artefact and individual innovation were what mattered. Space, following Kant, became conceived as a container with an explicit geometry that was filled with people and things, and cartography sought to represent that geometry. Scientific principles of collecting and mapping data emerged, but cartography was often seen as much of an art as a science, the product of the individual skill and eye of the cartographer. Mapping science was practical and applied and numerous small advances built a discipline.

In the latter part of the twentieth century, US scholar Arthur Robinson and his collaborators sought to re-cast cartography, focusing in particular on systematically detailing map design principles with the map user in mind. His aim was to create a science of cartography that would produce what he termed 'map effectiveness' – that is, maps that capture and portray relevant information in a way that the map reader can analyse and interpret (cf. Robinson and Petchenik 1976). Robinson suggested that an instrumental approach to mapping grounded in experimental psychology might be the best way for cartography to gain intellectual respectability and develop a rigorously derived and empirically tested body of generalizations appropriate for growing the new subject scientifically. Robinson adopted a view of the mind as an information-processing device. Drawing upon Claude Shannon's work in information theory, complexity of meaning was simplified into an approach focusing on input, transfer and output of information about the world. Social context was deemed to be irrelevant; the world existed independent of the observer and maps sought only to map the world. The cartographer was separate from the user and optimal maps could be produced to meet different needs.

The aims of the cartographer were normative – to reduce error in the representation and to increase map effectiveness through good design. Research thus sought to improve map designs by carefully controlled scientific experimentation that focused on issues such as how to represent location, direction and distance; how to select information; how best to symbolize these data; how to combine these symbols together; and what kind of map to publish. Framed by an empiricist ideology, the research agenda of cartography then was to reduce signal distortion in the communication of data to users. Art and beauty had no place in this functional cartographic universe.

Out of this context in the late 1960s and 1970s emerged an increasingly sophisticated series of attempts to develop and position cartographic communication models as the dominant theoretical framework to direct academic research. Communication models encouraged researchers to look beyond a functional analysis of map design, exploring filters that might hinder the encoding and decoding of spatial information (Figure 1.1). For researchers such as Grant Head (1984) or Hansgeorg Schlichtmann (1979) the map artefact became the focus of study, with an emphasis on the semiotic power of the map as opposed to its functional capacity, while Christopher Board (1981) showed how the map could be conceived as a conceptual, as well as a functional, model of the world. As models of cartographic communication multiplied so attention also increasingly focused on the map reader, with cognitive research seeking to understand how maps worked, in the sense of how readers interpreted and employed the knowledge maps sought to convey. Drawing on behavioural geography, it was assumed that map reading depended in large part upon cognitive structures and processes and research sought to understand how people came to know the world around them and how they made choices and decisions based on that knowledge. This approach is exemplified in the work of Reginald Golledge (1999), Robert Lloyd (2005) and Cynthia Brewer (cf. Brewer *et al.* 1997). Here the map user is conceived as an apolitical recipient of knowledge and the cartographer as a technician striving to deliver spatially precise, value-free representations that were the product of carefully controlled laboratory-based experiments that gradually and incrementally improved cartographic knowledge and praxis. Most research investigated the filters in the centre of this system concerned with the cartographers' design practice, and the initial stages of readers extracting information from the map (such work continues, e.g. Fabrikant *et al.* 2008). Little work addressed either what should be mapped or how mapping was employed socially because this was beyond the philosophical remit for valid research.

Other strands of scientific research into mapping emphasized the technologies that might be employed. Waldo Tobler's (1976) analytical cartography emerged in the early 1970s, offering a purely mathematical way of knowing the world, and laying the foundations for the emergence of

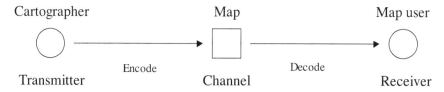

Figure 1.1 The basic map communication model, conceptualizing cartography in terms of stages in the transmission of spatial data from cartographer to reader via the map. Source: redrawn from Keates 1996: 114.

geographic information science. This analytical approach sought progress through the application of mathematical models and the subsequent application of technology so as to create new conceptual bases for mapping the world. Over time, conceptual and technically driven developments in computer graphics, computation and user interfaces have begun to fundamentally transmute the role of the map from a finished product to a situation where the map is displayed within a visual toolbox to be used interactively for exploratory data analysis (typically with the interlinking of multiple representations such as statistical charts, three-dimensional plots, tables and so on). This changing conceptualization of the map is at the heart of the emerging field of geovisualization, which in the last decade or so has been one of the leading areas of applied cartographic research (cf. Dodge *et al.* 2008; Dykes *et al.* 2005). Although distinctly positivist epistemologies underlie most of the geovisualization research, some have tried to open up the scope of visualization in more politically progressive directions, for example, Craine and Aitken's chapter in this volume, which considers the emotional energy latent in cinematic qualities of maps, and Kwan's (2007) work in fusing geospatial technologies with feminist theory to map affect and emotional geographies.

In other contexts different theoretical positions were adopted. For example, the French disciplinary tradition was much less influenced by Robinsonian functionalism and empirical research. Semiotic approaches were much more influential in this context, and may be traced back to the influential theories of Jacques Bertin. In 1967 Bertin derived from first principles a set of visual variables that might be manipulated by designers concerned with the effective design of mapping and other visualizations.

By the mid 1980s the cartographic communication model as an organizing framework for academic research was beginning to wane. Technological changes rendered problematic a single authoritative view of the world at a time when data were becoming much more readily available, and when technologies for the manipulation and dissemination of mapping were also being significantly changed. Users could become mappers and many possible mappings could be made. Digital mapping technologies separated display from printing and removed the constraint of fixed specifications. GIS increasingly supplanted many technical aspects of cartographic compilation and production. Digital position, elevation and attribute data could be captured from remotely sensed sources, and easily stored and manipulated in a digital form. Imagery could be generated to provide frequent updates of changing contexts. Maps could become animated. From the late 1990s the Internet has allowed maps to be evermore widely shared and disseminated at low cost. Mapping needed to be understood as much more of a process than was possible in communication models.

In the face of these profound challenges a second dominant approach to mapping research had replaced cartographic communication by the mid 1990s as the scientific orthodoxy. The linear inevitability of communication was supplanted by a multifaceted and multilayered merging of cognitive and semiotic approaches, centred on representational theory, and strongly influenced by the work of Alan MacEachren (1995). Articulating ideas grounded in Peircean semiotics, this approach recognized the need for a much less literal and functional positioning of maps. The iconic diagrammatic description of this approach is the notion of 'cartography cubed' (Figure 1.2). The dimensions of interactivity, the kind of knowledge, and the social nature of the process show the three key ways in which scientific understanding has been repositioned. Mapping can now be investigated as collaborative, the social context beyond map reading per se can be charted, and the process of knowing explored. And mapping is one of many kinds of visualization. However, mapping is still about revealing truth through a scientific approach reliant upon Western ways of seeing and upon technologies of vision; it still depends upon scientific experimentation and a representational view of the world.

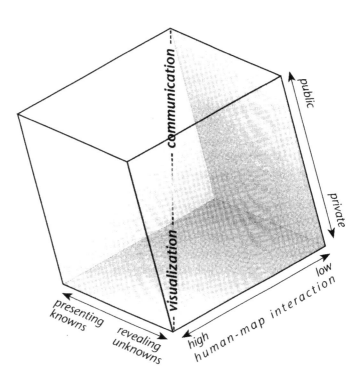

Figure 1.2 MacEachren's conceptual device, the 'cartography cube' employs the three different axes to encapsulate the distinctive characteristics of contemporary map use. Source: MacEachren 1994: 6.

Maps as social constructions

The view that cartography produces maps of truth in an objective, neutral, scientific fashion has been challenged by a number of scholars. In the late 1980s, the work of Brian Harley began to question how mapping operated as a powerful discourse, challenging the scientific orthodoxy of cartographic research. He proposed a new research agenda concerned with the roles maps play in different societies, arguing that maps often reinforce the status quo or the interests of the powerful, and that we should investigate the historical and social context in which mapping has been employed. In this view cartography was not necessarily what cartographers said it was. Instead, Harley argued that we could only understand the history of cartography if we interrogate the forces at play around mapping.

Harley (1989) drew on the ideas of Michel Foucault among others to argue that the process of mapping was not a neutral, objective pursuit but rather was one laden with power. He contended that the process of mapping consists of creating, rather than simply revealing, knowledge. In the process of creation many subjective decisions are made about what to include, how the map will look and what the map is seeking to communicate. As such, Harley noted, maps are imbued with the values and judgements of the individuals who construct them and they are undeniably a reflection of the culture in which those individuals live. Maps are typically the products of privileged and formalized knowledges and they also tend to produce certain kinds of knowledge about the world. And in this sense, maps are the products of power and they produce power. In contrast to the scientific view that positions maps in essentialist terms, Harley cast maps as social constructions; as expressions of power/knowledge. Others, such as Denis Wood (1992) and John Pickles (2004), have extensively demonstrated this power/knowledge revealing the ideology inherent in maps (or their 'second text') and how maps 'lie' (or at least provide selective stories while denying their selectivity) due to the choices and decisions that have to be made during their creation, and through how they are read by users.

This social constructivist critique sometimes also articulated structural explanations for mapping, which sought understanding beneath the apparent surface of observable evidence. For example, David Harvey's (1989) Marxist analysis of the role of mapping in time–space compression examined the role of global images in the expansion of European colonial powers, and situated these as reflections of a changing mode of production. Drawing on linguistic structural thought Denis Wood (1992) employed Barthean semiotics to persuasively argue that the power of maps lay in the interests they represented. Mapping in this view always has a political purpose, and this 'interest' often leads to people being pushed 'off the map'. Wood argued that mapping works through a shared cultural reading of a number of different codes in every map, which may be analysed in a semiotic process to reveal the power behind the map. These interests all too often led to subjugation, oppression, control and inequality. Through economic relations, legal evidence, governance or social practice the power of maps continues to be used to control. It has been argued that many of the social roles played by cartographic knowledge stem from the modernist project, and that a mapping mentality is integral to the modernist enterprise itself (Cosgrove and Martins 2000). By examining different categories across which power might be articulated contextual studies can reveal how maps reflect but also constitute different kinds of political relation. Colonialism, property ownership, national identity, race, military power, bureaucracy and gender have all been theorized as playing key roles in mapping relations (see Anderson 1991; Haraway 1992; Pickles 2004).

For example, local knowledge has been translated into tools to serve the needs of the colonizer, with new territories scripted as blank spaces, empty and available for the civilizing Western explorer to claim, name, subjugate and colonize (Edney 1997). Projection and design have been used to naturalize the political process of imperial control and sell imperial values to citizens at home. The continuing progress of colonial adventures is mapped out nowadays in our news broadcasts and on the Internet, but the imperial rhetoric of control, governance, management of territory and creation of new imperial landscapes remains the same (cf. Gregory 2004). The colonial project relies on the map, and in turn the map relies on colonial aspirations.

The work by Harley, Wood, Harvey and others set the groundwork for work since the 1990s that has been labelled critical cartography (see Crampton and Krygier 2005) and with respect to wider geospatial technologies, critical GIS (see O'Sullivan 2006; Schuurman 1999). Critical cartography is avowedly political in its analysis of mapping praxis seeking to deconstruct the work of spatial representations in the world and the science that produces them. It is, however, decidedly not against maps, but rather seeks to appreciate the diverse ways in which maps are produced and used by different individuals and groups. From such a perspective there is no one 'right way' to produce maps, but their makers need to be sensitive to politics and context of their making and use. For some theorists this means moving beyond thinking of maps as representations to try to conceive of a post-representational cartography.

Post-representational cartography

From ontic knowledge to ontology

Despite the obvious advances of the various social constructivist approaches in rethinking maps, more recent work has sought to further refine cartographic thought and to construct post-representational theories of mapping. Here, scholars are concerned that the critique developed by Harley and others did not go far enough in rethinking the ontological bases for cartography, which for them has too long been straitjacketed by representational thinking. As

Denis Wood (1993) and Jeremy Crampton (2003) outline, Harley's application of Foucault to cartography is limited. Harley's observations, although opening a new view onto cartography, stopped short of following Foucault's line of inquiry to its logical conclusion. Instead, Crampton (2003: 7) argues that Harley's writings 'remained mired in the modernist conception of maps as documents charged with "confessing" the truth of the landscape'. In other words, Harley believed that the truth of the landscape could still be revealed if one took account of the ideology inherent in the representation. The problem was not the map per se, but 'the bad things people *did* with maps' (Wood 1993: 50, original emphasis); the map conveys an inherent truth as the map remains ideologically neutral, with ideology bound to the subject of the map and not the map itself. Harley's strategy was then to identify the politics of representation in order to circumnavigate them (to reveal the truth lurking underneath), not fully appreciating, as with Foucault's observations, that there is no escaping the entangling of power/knowledge.

Crampton's solution to the limitations of Harley's social constructivist thinking is to extend the use of Foucault and to draw on the ideas of Heidegger and other critical cartographers such as Edney (1993). In short, Crampton (2003: 7) outlines a 'non-confessional understanding of spatial representation' wherein maps instead of 'being interpreted as objects at a distance from the world, regarding that world from nowhere, that they be understood as being in the world, as open to the disclosure of things'. Such a shift, Crampton argues, necessitates a move from understanding cartography as a set of ontic knowledges to examining its ontological terms. Ontic knowledge consists of the examination of how a topic should proceed from within its own framework where the ontological assumptions about how the world can be known and measured are implicitly secure and beyond doubt (Crampton 2003). In other words, there is a core foundational knowledge – a taken for granted ontology – that unquestioningly underpins ontic knowledge.

With respect to cartography this foundational ontology is that the world can be objectively and truthfully mapped using scientific techniques that capture and display spatial information. Cartography in these terms is purely technical and develops by asking self-referential, procedural questions of itself that aim to refine and improve how maps are designed and communicate (Crampton gives the examples of what colour scheme to use, the effects of scale, how maps are used historically and politically). In these terms a book like Robinson *et al.* (1995) is a technical manual that does not question the ontological assumptions of the form of mapping advocated, rather it is a 'how to do "proper" cartography' book that in itself perpetuates the security of cartography's ontic knowledge. In this sense, Harley's questioning of maps is also ontical (e.g. see Harley 1992), as his project sought to highlight the ideology inherent in maps (and thus expose the truth hidden underneath) rather than to question the project of mapping per se; 'it provided an epistemological avenue into the map, but still left open the question of the ontology of the map' (Crampton 2003: 90). In contrast, Crampton details that examining cartography ontologically consists of questioning the project of cartography itself.

Such a view leads to Crampton, following Edney (1993), to argue for the development of a non-progressivist history of cartography; the development of a historical ontology that rather than being teleological (wherein a monolithic view of the history of cartographic practices is adopted that sees cartography on a single path leading to more and more complete, accurate and truthful maps) is contingent and relational (wherein mapping – and truth – is seen as contingent on the social, cultural and technical relations at particular times and places). Maps from this perspective are historical products operating within 'a certain horizon of possibilities' (Crampton 2003: 51). (See also his chapter in this volume that discusses the ways different forms of mapping inframe racial identities with important ramifications for how humanity is made visible.) It thus follows that maps created in the present are products of the here-and-now, no better than maps of previous generations, but rather different to them. Defining a map is dependent on when and where the map was created, as what constitutes a map has changed over time. For Crampton (2003: 51) this means that a politics of mapping should move beyond a 'critique of existing maps' to consist of 'a more sweeping project of examining and breaking through the boundaries on how maps are, and our projects and practices with them'; it is about exploring the 'being of maps'; how maps are conceptually framed in order to make sense of the world. Several other cartographic theorists have been following similar lines of enquiry to Crampton in seeking to transfer map theory from ontic knowledge to ontology and it is to them that we now turn.

Maps as inscriptions

John Pickles (2004) has sought to extend cartographic theory beyond ontic status by conceiving of maps as inscriptions as opposed to representations or constructions. His work focuses on 'the work that maps do, how they act to shape our understanding of the world, and how they code that world' (p. 12). As such his aim is to chart the 'practices, institutions and discourses' of maps and their social roles within historical, social and political contexts using a poststructural framework that understands maps as complex, multivocal and contested, and which rejects the notion of some 'truth' that can be uncovered by exposing ideological intent. Pickles' detailed argument unpicks the science of representation, calling for a post-representational cartography that understands maps not as mirrors of nature, but as producers of nature. To paraphrase Heisenberg (1959, cited in Pickles 2004), Pickles argues that cartography does not simply describe and explain the world; it is part of the interplay between the world and ourselves; it describes the world as exposed to our method of questioning.

For Pickles, maps work neither denotatively (shaped by the cartographic representation, labelling, embedded with other material such as explanatory text, etc.) or connotatively (what the mapper brings to the representation in terms of skills, knowledges, etc.) but as a fusion of the two. Pickles thus proposes a hermeneutic approach that interprets maps as unstable and complex texts, texts that are not authored or read in simple ways. Rather than a determinate reading of the power of maps that seeks to uncover in a literal sense the authorial and ideological intent of a map (who made the map and for what purpose), Pickles expresses caution in fixing responsibility in such a manner, recognizing the multiple, institutional and contextual nature of mapping. Similarly, the power of maps is diffuse, reliant on actors embedded in contexts to mobilize their *potential* effects: 'All texts are . . . embedded within chains of signification: meaning is dialogic, polyphonic and multivocal – open to, and demanding of us, a process of ceaseless contextualization and recontextualization' (Pickles 2004: 174).

Alongside a hermeneutic analysis of maps, Pickles proposes that a post-representational cartography consists of the writing of denaturalized histories of cartography and the production of de-ontologized cartography. Denatural-ized histories reveal the historicizing and contextualizing conditions that have shaped cartographic practices to 'explore the ways in which particular machines, disciplines, styles of reasoning, conceptual systems, bodies of knowledge, social actors of different scales . . . and so forth, have been aligned at particular times and particular places' (Pickering 1995, quoted in Pickles 2004: 70). In other words, they consist of genealogies of how cartography has been naturalized and institutionalized across space and time as particular forms of scientific practices and knowledge. A de-ontologized cartography is on the one hand about accepting counter-mappings as having equal ontological status as scientific cartographic (that there are many valid, cartographic ontologies), and on the other, deconstructing, reading differently, and reconfiguring scientific cartography (to examine alternative and new forms of mapping).

Maps as propositions

Like Pickles, Crampton and others, Wood and Fels (2008) extend the notion of a map as social construction to argue that the map itself, its very make-up and construction – its self-presentation and design, its symbol set and categorisation, its attendant text and supporting discourse – is ideologically loaded to convey particular messages. A map does not simply represent the world; it produces the world. They argue that maps produce the world by making propositions that are placed in the space of the map. Maps achieve their work by exclaiming such propositions and Wood and Fels define this process as one of 'posting' information on map. Posting is the means by which an attribute is recognized as valid (e.g. some class of the natural world) and is spatialized. It is the means by which the *nature* of maps (is – category) and the nature of *maps* (there – sign) conjoin to create a unified spatial ontology (this is there). However, the map extends beyond spatial ontology by enabling higher order propositions (this is there and *therefore it is also*; Wood and Fels 2008) to link things in places into a relational grid.

Wood and Fels argue that the power of this spatial propositional framework is affirmed through its call to authority – by being an objective reference object that is prescriptive not descriptive. So the map produces and reaffirms territory rather than just describing it. Authority is conveyed through what they term the paramap. A paramap is the combination of perimap and epimap. The perimap consists of the production surrounding a map: the quality of the paper, the professionalism of the design, the title, legend, scale, cartouches, its presentation and so on. The epimap consists of the discourse circulating a map designed to shape its reception: advertisements, letters to reviewers, endorsements, lectures, articles, etc. Together, the perimap and epimap work to position the map in a certain way and to lend it the authority to do work in the world.

Because maps are prescriptive systems of propositions, Wood and Fels contend that map creation should not solely be about presenting information through attractive spatial representations as advocated by the majority of cartographic textbooks (which borrow heavily from graphic design traditions). Instead they suggest map design should be about the 'construction of meaning as a basis for action' (p. 14). They propose turning to cognitive linguistics to rethink map design as a form of 'cognitive cartographics'. Cognitive linguistics examines the ways in which words activate neural assemblages and open up 'thinking spaces' in the mind within which meaning is constructed by linking present information with past knowledge. They contend that maps perform like words, by firing up thinking spaces. Employing cognitive cartographics, they suggest, will create a non-representational approach to map design focused on the construction of meaning rather than graphic design and the nature of signs. It will also enable cartographic theory to move beyond the compartmentalized thinking that has divided mapmaking from map use by providing a more holistic framework. In other words, both map design and map reading can be understood through a cognitive cartographics framework. These ideas are developed in Krygier and Wood's chapter in this volume.

Maps as immutable mobiles and actants

In his book *Science in Action* Bruno Latour (1987) used the example of cartography to explore how the cultures and mechanisms involved in production of Western scientific knowledge gained their power and authority to make truth claims about the world that in turn are employed to do work in the world. He cogently argued that the assemblage of cartographic theory, mapping technologies (e.g. quadrants, sextants, log books, marine clocks, rulers, etc.), and disciplinary regimes of trade and service (e.g. sea captains all taught the same principles and practices of surveying, recording and bringing back spatial data) worked together to enable information from distant places to be accumulated in a cyclical and systematic fashion and for maps to enable appropriate action at a distance (maps informed their readers as to local conditions and guided their safe navigation).

As the scientific basis of mapmaking and map use became conventionalized, Latour argues that maps increasingly took on the status of immutable mobiles. That is, the mechanisms used to generate cartographic information and the form maps took (in terms of scale, legend, symbols, projection, etc.) became familiar and standardized through protocols so that the map became a stable, combinable and transferable form of knowledge that is portable across space and time. As such, a map produced in South America by Argentinian cartographers is decipherable to someone from another country because it shares common principles that render it legible. Moreover, spatial data transported from South America in the form of latitude and longitude can be used to update charts of the area or be combined with other information, despite the fact that the cartographer is unlikely to have ever visited the area they are mapping.

Mapping then is seemingly transformed into a 'universal' scientific practice and maps become mobile and immutable artefacts through which the world can be known and a vehicle through which spatial knowledge can be transported into new contexts. What is mapped, how it is mapped, and the power of maps is the result of Western science's ability to set the parameters and to dominate the debate about legitimate forms of knowledge. As Latour notes, however, cartographic theory and praxis is seemingly immutable in nature because it disciplines its practitioners and silences other local mapping knowledges. And yet, immutable Western cartographic practice is itself similarly the product of localized practices that are deemed appropriate within a limited circle of practitioners and mapping agencies, who exercise powerful claims to scientific objectivity and truth. The immutability of maps is then at one level a powerful illusion, but one that readily does work in the world.

Latour contends that the immutability, combinability and mobility of maps allowed exploration, trade and ultimately colonialism to develop by allowing control to be exerted from afar and knowledges about new territories to be effectively transported globally. Maps became a vital part in the cycle of knowledge accumulation that allowed explorers to '*bring* the lands *back with them*' and to successfully send others in their footsteps (Latour 1987: 220, original emphasis). Latour thus argues that the European cartographers of the Renaissance produced centres of calculation (key sites of cartographic practice) that came to dominate the world. In so doing, maps he suggests do not simply represent space at a particular time, but produce new spaces – times. Maps open up new possibilities – such as international trade and territorial conquest – and thus create new geographies and histories.

To understand maps then, Latour suggests that it is necessary to unpick the cultures, technologies and mechanics of how a particular form of mapping came to gain immutability and mobility to reveal its contingencies and relationalities. Following on from his work, the development of Actor-Network

Theory (ANT) in science studies has provided a framework for considering how maps work in concert with other actants and actors to transform the world. ANT involves the tracing out of the context and instruments of mapping – its assemblage – not just cartographic praxis. For example, understanding the road system, Latour argues, cannot be fully realized by looking at infrastructure and vehicles alone, it also needs to consider civil engineering, plans of roads, standards for signage, garages, mechanics, drivers, political lobbying, funding, spare parts and so on.

Maps do not have meaning or action on their own; they are part of an assemblage of people, discursive processes and material things. They are deployed in an actor-network of practices rather than existing as de-corporalized, a priori, non-ideological knowledge objects. ANT then seeks to provide a broader and richer understanding of the creation of maps through particular actor-networks (e.g. a national mapping agency) and the use of maps as actants within various actor-networks (e.g. land conservation) by considering the diverse, day-to-day practices of, and the interactions and the circulation of ideas and power between, various actors (people, texts, objects and money) (Perkins 2006). In so doing, ANT identifies the nature of 'boundary objects' (objects such as technical standards that enable the sharing of information across networks), 'centres of calculation' (locations such as mapping agencies where observations are accumulated, synthesized and analysed), 'inscription devices' (technical artefacts that record and translate information such as tables of coordinates or satellite imagery), 'obligatory points of passage' (a site in a network that exerts control and influence such as government department), 'programs of action' (the resources required for an actor to perform certain roles) and 'trials by strength' (how competing visions and processes within the network compete for superiority) (cf. Martin 2000). From this perspective, the stories of mapping always need to be considered as historically contingent actor-networks; as timed, placed, cultured and negotiated; a Web of interacting possibilities in which the world is complex and nothing is inevitable. The focus shifts from what the map represents to how it is produced and how it produces work in the world (Perkins 2006).

From ontology to ontogenesis: maps as practices

In recent years, there has been a move towards considering cartography from a relational perspective, treating maps not as unified representations but as constellations of ongoing processes. Here it is recognized that maps are produced and used through multiple sets of practices. Spatial data are surveyed, processed and cleaned; geometric shapes are drafted, revised, updated, copied, digitized and scanned; information is selected for inclusion, generalized and symbolized. A map is then worked upon by the world and does work in the world. It might be folded or rolled, converted to another file format, embedded in other media; it might be packaged, marketed, sold, bought, used, stored, collected, re-used, thrown away or recycled; it might be read in different ways in different contexts; it might be employed to plan a journey, make money, play a game (see Perkins in this volume) or teach moral values. Mapmaking and map use are understood as processual in nature, being both embodied and dynamic.

Mapping can then be conceptualized as a suite of cultural practices involving action and affects. This kind of approach reflects a philosophical shift towards performance and mobility and away from essence and material stability. This rethinking of cartography is supported by historical and contemporary work. Researchers concerned with historical contexts increasingly stress the interplay between place, times, actions and ideas. Mapping in different cultures reflects multiple traditions including an internal or cognitive set of behaviours involving thinking about space; a material culture in which mapping is recorded as an artefact or object; and a performance tradition where space may be enacted through gesture, ritual, song, speech dance or poetry (Woodward and Lewis 1998). In any cultural context there will be a different blend of these elements. Interpreting mapping then means considering the context in which mapping takes place; the way it is invoked as part of diverse practices to do work in the world. Instead of focusing on artefacts, aesthetics, human agency, or the politics of mapping, research focuses on how maps are constituted in and through diverse, discursive and material processes.

Arguments presently emerging in the literature extend both the notion of maps as processes and the ontological thought underpinning cartography by problematizing the ontological security enjoyed by maps. The idea that a map represents spatial truth might have been challenged and rethought in a number of different ways, but a map is nonetheless understood as a coherent, stable product – *a* map; a map has an undeniable essence that can be interrogated and from which one can derive understanding. Moreover, the maps and mapping practices maintain and reinforce dualities with respect to their conceptualization – production–consumption, author–reader, design–use, representation–practice, map–space. This position has been rejected by those adopting performative and ontogenetic understandings of mapping. Maps rather are understood as always in a state of becoming; as always mapping; as simultaneously being produced *and* consumed, authored *and* read, designed *and* used, serving as a representation *and* practice; as mutually constituting map/space in a dyadic relationship.

James Corner (1999) argues that cartographic theory has been hampered by a preoccupation to view maps in terms of what they represent and mean rather than what they do. Drawing on poststructural theory, he problematizes the conception of maps as representations that are separate and proceeding from territory. Following Baudrillard, Corner argues that a territory does not precede a map, but that space becomes territory through bounding practices that include mapping. Moreover, given that places are planned and built on the basis of maps, so that space is itself a representation of the map, the 'differentiation between the real and the representation is no longer meaningful' (p. 222). Maps and territories are co-constructed. Space is constituted through mapping practices, among many others, so that maps are not a reflection of the world, but a re-creation of it; mapping activates territory.

Corner develops an understanding of maps as unfolding potential; as conduits of possibilities; as the sites of imagination and action in the world. The 'function of maps is not to depict but to enable'; 'mappings do not *represent geographies* of ideas; rather they *effect* actualization' (p. 225, original emphasis). Mapping involves *processes* of 'gathering, working, reworking, assembling, relating, sifting, . . . speculating and so on . . . [that] allow certain sets of possibility to become actual' (p. 228, our emphasis). In this sense, maps remake 'territory over and over again, each time with new and diverse consequences' (p. 213). Corner explains that maps engender such re-territorializations because they are doubly projective: they both capture elements from the world and also project back a variety of effects through their use. As such, the agency of maps lies not in 'their reproduction or imposition, but in uncovering realities previously unseen or unimagined' (p. 213). He thus suggests that cartographic research and practice need to focus on mapping actions and mapping effects and not solely on the construction of maps per se. He charts four practices of mapping – drift, layering, game-space and rhizome – to illustrate how the processes of mapping and the ongoing construction of space entwine. To take one of these, Corner (1999: 244) argues that the map acts as a rhizome because it is infinitely open with many diverse entry points and exits that enable 'a plurality of readings, uses and effects', opening up milieus to new possibilities of action. So a 'standard' topographic map sheet from the Ordnance Survey, for example, has 'multiple entryways, diverse uses and applications, infinite routes and networks, and potentially endless surfaces of engagement' (p. 246) that when enacted brings the world into being in new ways.

Tim Ingold (2000) also develops an approach to mapping grounded in cultural practice. He makes a distinction between mapping, mapmaking and map-use and argues that map-use (navigation) is to navigate by means of a map, plotting a course between one location to another in space. Mapping, in terms of wayfinding practices, however, consists of moving from one place to another in a region. He argues that maps that chart peoples' experiences of movement – such as sketch maps, indigenous maps – are expressions of mapping. For him, because these mappings refer to the itineraries of their inhabitants they do not detail locations in space but histories of movement that constitute place. Such movements consist of passages through vistas, rather than an abstracted Cartesian landscape and therefore encode mobility as opposed to location (see Figure 1.3). As such, the resulting mappings are 'not so much representations of space as condensed histories' (Ingold 2000: 220) and therefore un-maplike. They are un-maplike because the knowledge they portray is bound to the place where they are made, unlike Western cartographic practice, which seeks to be non-indexical – that is a view from nowhere. However, as Turnbull (1989) and others have noted,

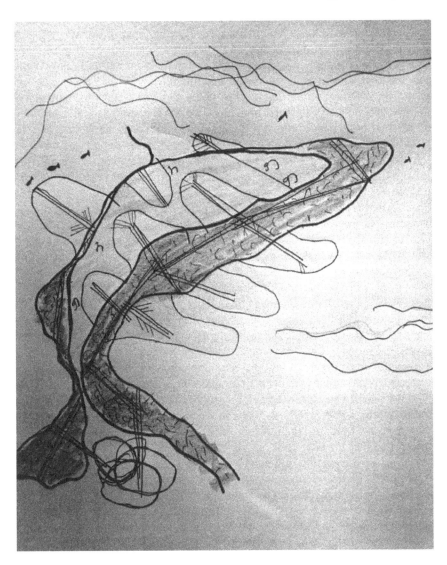

Figure 1.3 A paper rendering of indigenous hunting 'map' created by an Andamanese person for an anthropology researcher. Source: Pandya 1990: 790.

the non-indexical nature of maps is an illusion – they are always a view from somewhere bound within the practices and knowledges of their makers.

Western cartography, according to Ingold (2000: 203), thus 'transforms everywhere-as-region, the world as experienced by a mobile inhabitant, into everywhere-as-space, the imaginary "bird's-eye view" of a transcendent consciousness' (see also Propen's chapter in this volume in which he discusses the nature of disembodied views of the whole earth). In so doing, people and their experiences are obliterated from the map, and the structure of the world is fixed without regard to the movements and actions of its inhabitants – 'the world it describes is not a world in the making, but one ready-made for life to occupy' (p. 235); 'in the cartographic world . . . all is still and silent' (p. 242). Maps as reminders of paths and expressions of experience, as they were conceived in the European Middle Ages, morphed into supposed representations of space through the application of scientific principles. The issue is, however, that people live in the everywhere-as-region and know as they go – they are constantly mapping as they move through places employing a form of process cartography – so there is a disconnect between Western notions of a map, and the everyday ways in which people come to know and be in the world. This leads to a paradox – the more a map 'aims to furnish a precise and comprehensive representation of reality, the less true to life this representation appears' (p. 242). For Ingold, we need to simultaneously understand and value the process cartography of mapping and critique and reform representational modes of cartography.

Vincent Del Casino and Stephen Hanna (2005) draw on poststructural theory, and in particular the ideas of Deleuze and Guattari and Judith Butler, to argue that maps are in a constant state of becoming; that they are 'mobile subjects' whose meaning emerges through socio-spatial practices of use that mutate with context and is contested and intertextual. For them the map is not fixed at the moment of initial construction, but is in constant modification where each encounter with the map produces new meanings and engagements with the world. Del Casino and Hanna (2005: 36) state that '[m]aps are both representations and practices . . . simultaneously. Neither is fully inscribed with meaning as representations nor fully acted out as practices'. In so doing, they argue that maps are not 'simply visual objects ripe for deconstruction. . . . Maps . . . are tactile, olfactory, sensed objects/subjects mediated by the multiplicity of knowledges we bring to and take from them through our everyday interactions and representational and discursive practices' (p. 37).

Maps and spaces co-produce each other through spatial practices to create what they term 'map spaces', wherein it is impossible to disentangle fully how the map does work in the world from how the world shapes how the map is performed – they are co-constitutive. Del Casino and Hanna (2005) illustrate their arguments by an examination of how visitors produce the historic town of Fredericksburg in Virginia, by deploying tourist maps, along with other texts and narratives (such as a guided tour), which together shape how people interact with the space and the town. They show that the real is read back into the map, making it more legible. Tourists are both consumers and producers of the map; authors and readers. Meaning emerges through action and action is shaped by meaning in a complex, recursive and intertextual performativity. The tourist map of Fredericksburg then is never complete, but is always mobile; always being produced by tourists and producing Fredericksburg.

In a similar vein, Rob Kitchin and Martin Dodge (2007) have argued that map theory needs to shift in perspective from seeking to understand the nature of maps (how maps are) to examining the practices of mapping (how maps become). Maps they argue are not ontologically secure representations but rather a set of unfolding practices. They state:

> [m]aps are of-the-moment, brought into being through practices (embodied, social, technical), *always* re-made every time they are engaged with; mapping is a process of constant re-territorialization. As such, maps are transitory and fleeting, being contingent, relational and context-dependent. *Maps are practices* – they are always *mappings*; spatial practices enacted to solve relational problems (e.g. how best to create a spatial representation, how to understand a spatial distribution, how to get between A and B, and so on).
>
> (Kitchin and Dodge 2007: 5, original emphasis)

From this perspective, they contended that Figure 1.4 is not unquestioningly a map; it is rather a set of points, lines and colours that is brought into being as a map through mapping practices (an inscription in a constant state of re-inscription). As such, the map is (re)made *every time* mapping practices, such as recognizing, interpreting, translating and communicating, are applied to the pattern of ink. These mapping practices give the map the semblance of an immutable mobile and ontological security because they are learned and constantly reaffirmed. As Pickles explains:

> [m]aps work by naturalizing themselves by reproducing a particular sign system and at the same time treating that sign system as natural and given. But, map knowledge is never naïvely given. It has to be learned and the mapping codes and skills have to be culturally reproduced.
>
> (2004: 60–1)

Maps do not then emerge in the same way for all individuals. Rather they emerge in contexts and through a mix of creative, reflexive, playful, tactile and habitual practices, affected by the knowledge, experience and skill of the individual to perform mappings and apply them in the world. This applies as much for mapmaking as for map reading. As such, the map does not re-present the world or make the world: it is a co-constitutive production between inscription, individual and world; a production that is constantly in motion, always seeking to appear ontologically secure. Conceiving of maps in this way reveals that they are never fully formed but emerge in process and are mutable (they are re-made as opposed to mis-made, mis-used or mis-read).

In terms of cartographic research, this conceptualization of maps necessitates an epistemology that concentrates on how maps emerge – how maps are made through the practices of the cartographer situated within particular contexts and how maps re-make the world through mutually constituted

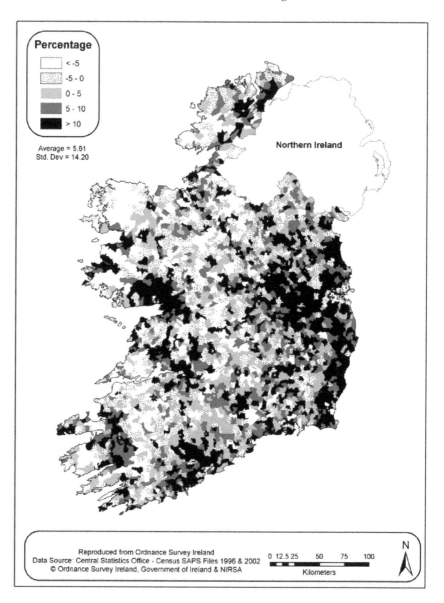

Figure 1.4 Is this image a map? Population change in Ireland, 1996–2002. Source: R. Kitchin.

practices that unite map and space. As Brown and Laurier (2005: 19, original emphasis) note, this requires a radical shift in approach from '*imagined* scenarios, *controlled* experiments or *retrospective* accounts' to examine how maps emerge as solutions to relational problems; to make sense of the 'unfolding action' of mapping. Their approach is the production of detailed ethnographies of how maps become; mapmaking and use are observed in specific, local contexts to understand the ways in which maps are constructed and embedded within cultures of practices and affect. In their study they examined how maps are used in the context of navigating while driving between locations through video-based ethnography. Their work highlighted how a map, journey and social interaction within the car emerged through each other in contingent and relational ways within the context of the trip.

Conclusion

Mapping, its theory, praxis and technologies, is a rapidly changing and exciting field of study. Intellect, capital, culture and innovation are reshaping how maps are made, used and thought about. In this book we are particularly concerned with exploring the diverse constellation of contemporary mapping theories. As we have so far demonstrated, the theories of mapping consist of a set of winding and contested journeys through philosophical and practical terrains. These journeys are far from over and the philosophical under-pinnings of maps remain a fertile ground in which to explore issues of space, representation and praxis. The chapters that follow provide detailed examinations into contemporary cartographic theory. They highlight that there are many rich ways of rethinking maps both ontologically and epistemo-logically. It is certainly not clear if any of these different modes of thought will emerge to become paradigmatic and it may be the case that we are entering a period characterized by theoretical diversity and exchange. For us, such a period will continue to be highly productive in terms of thinking through the nature and role of maps, their production and use, and the work that they do in the world. There is much rethinking yet to be done!

Note

1   Parts of this chapter draw upon material from Kitchin and Dodge (2007), Kitchin (2008) and Perkins (2009a and 2009b).

References

Anderson, B. (1991) *Imagined Communities: Reflections on the Origin and Spread of Nationalism*, London: Verso.
Bertin, J. (1967) *Sémiologie Graphique*, Paris: Gauthier-Villars.
Board, C. (1981) 'Cartographic communication', *Cartographica*, 18(2): 42–78.
Brewer, C.A., MacEachren, A.M., Pickle, L.W. and Herrmann, D. (1997) 'Mapping mortality: Evaluating color schemes for choropleth maps', *Annals of the Association of American Geographers*, 87(3): 411–38.
Brown, B. and Laurier, E. (2005) 'Maps and journeys: an ethno-methodological investigation', *Cartographica*, 40(3): 17–33.
Corner, J. (1999) 'The agency of mapping: speculation, critique and invention', in D. Cosgrove (ed.) *Mappings*, London: Reaktion Books.

Cosgrove, D. and Martins, L.L. (2000) 'Millennial geographics', *Annals of the Association of American Geographers*, 90(1): 97–113.

Crampton, J. (2003) *The Political Mapping of Cyberspace*, Edinburgh: Edinburgh University Press.

Crampton, J. and Krygier, J. (2005) 'An introduction to critical cartography', A*CME: An International E-Journal for Critical Geographies*, 4(1): 11–33.

Del Casino, V.J. and Hanna, S.P. (2005) 'Beyond the "binaries": A methodological intervention for interrogating maps as representational practices', *ACME: An International E-Journal for Critical Geographies*, 4(1): 34–56.

Dodge, M., McDerby, M. and Turner, M. (2008) *Geographic Visualization: Concepts, Tools and Applications*, Chichester, England: John Wiley & Sons.

Dykes, J., MacEachren, A.M. and Kraak, M-J. (2005) *Exploring Geovisualization*, London: Elsevier.

Edney, M.H. (1993) 'Cartography without "progress": Reinterpreting the nature and historical development of map-making', *Cartographica*, 30(2/3): 54–68.

Edney, M.H. (1997) *Mapping an Empire: The Geographical Construction of British India*, Chicago, IL: University of Chicago Press.

Fabrikant, S.I., Rebich-Hespanha, S., Andrienko, N., Andrienko, G. and Montello, D.R. (2008) 'Novel method to measure inference affordance in static small-multiple map displays representing dynamic processes', *The Cartographic Journal*, 45(3): 201–15.

Golledge, R.G. (1999) *Wayfinding Behavior: Cognitive Mapping and Other Spatial Processes*, Baltimore, MD: Johns Hopkins University Press.

Gregory, D. (2004) *The Colonial Present: Afghanistan, Palestine, Iraq*, London: Blackwell.

Haraway, D. (1992) *Simians, Cyborgs, and Women: The Reinvention of Nature*, New York: Routledge.

Harley, J.B. (1989) 'Deconstructing the map', *Cartographica*, 26(2): 1–20.

Harley, J.B. (1992) 'Rereading the maps of Columbian encounters', *Annals of the Association of American Geographers*, 82(3): 522–36.

Harvey, D. (1989) *The Condition of Postmodernity*, London: Blackwell.

Head, C.G. (1984) 'The map as natural language: a paradigm for understanding', *Cartographica*, 31(1): 1–32.

Henrikson, A.K. (1994) 'The power and politics of maps', in G.J. Demko and W.B. Wood (eds) *Reordering the World: Geopolitical Perspectives on the Twenty-First Century*, Boulder, CO: Westview Press.

Ingold, T. (2000) *The Perception of the Environment: Essays in Livelihood, Dwelling and Skill*, London: Routledge.

Keates, J.S. (1996) *Understand Maps*, Harlow, England: Addison Wesley.

Kitchin, R. (2008) 'The practices of mapping', *Cartographica*, 43(3): 211–15.

Kitchin, R. and Dodge, M. (2007) 'Rethinking maps', *Progress in Human Geography*, 31(3): 331–44.

Kwan, M-P. (2007) 'Affecting geospatial technologies: Toward a feminist politics of emotion', *The Professional Geographer*, 59(1): 22–34.

Latour, B. (1987) Science in Action, Cambridge, MA: Harvard University Press.

Lloyd, R. (2005) 'Assessment of simulated cognitive maps: The influence of prior knowledge from cartographic maps', *Cartography and Geographic Information Science,* 32: 161–79.

MacEachren, A.M. (1994) 'Visualization in modern cartography: setting the agenda', in A.M. MacEachren and D.R.F. Taylor (eds) *Visualization in Modern Cartography*, Oxford: Pergamon.

MacEachren, A.M. (1995) *How Maps Work: Representation, Visualization and Design*, New York: Guilford Press.

Martin, E. (2000) 'Actor-networks and implementation: Examples from conservation GIS in Ecuador', *International Journal of Geographical Information Science*, 14(8): 715–38.

O'Sullivan, D. (2006) 'Geographic information science: Critical GIS', *Progress in Human Geography*, 30(6): 783–91.

Pandya, V. (1990) 'Movement and space: Andamanese cartography', *American Ethnologist*, 17(4): 775–97.

Perkins, C. (2006) 'Mapping golf: A contextual study', *The Cartographic Journal*, 43(3): 208–23.

Perkins, C. (2009a) 'Performative and embodied mapping', in R. Kitchin and N. Thrift (eds) *International Encyclopedia of Human Geography*, Oxford: Elsevier.

Perkins, C. (2009b) 'Philosophy and mapping', in R. Kitchin and N. Thrift (eds) *International Encyclopedia of Human Geography*, Oxford: Elsevier.

Pickles, J. (2004) *A History of Spaces: Cartographic Reason, Mapping and the Geo-Coded World*, London: Routledge.

Raisz, E. (1938) *General Cartography*, New York: McGraw-Hill.

Robinson, A.H. and Petchenik, B.B. (1976) *The Nature of Maps*, Chicago, IL: University of Chicago Press.

Robinson, A.H., Morrison, J.L., Muehrcke, P.C., Kimmerling, A.J. and Guptil, S.C. (1995) *Elements of Cartography*, 6th edn, New York: Wiley.

Schlichtmann, H. (1979) 'Codes in map communication', *The Canadian Cartographer*, 16(1): 81–97.

Schuurman, N. (1999) 'Critical GIS: Theorizing an emerging discipline', *Cartographica*, 36(4): 5–21.

Tobler, W.R. (1976) 'Analytical cartography', *American Cartographer*, 3(1): 21–31.

Turnbull, D. (1989) *Maps are Territories: Science is an Atlas*, Chicago, IL: University of Chicago Press.

Wood, D. (1992) *The Power of Maps*, New York: Guilford Press.

Wood, D. (1993) 'The fine line between mapping and map-making', *Cartographica*, 30(4): 50–60.

Wood, D. and Fels, J. (2008) *The Natures of Maps: Cartographic Constructions of the Natural World*, Chicago, IL: University of Chicago Press.

Woodward, D. and Lewis, G.M. (1998) *The History of Cartography. Vol. 2. Book 3. Cartography in the Traditional African, American, Arctic, Australian, and Pacific Societies*, Chicago, IL: University of Chicago Press.

2 Rethinking maps and identity
Choropleths, clines, and biopolitics

Jeremy W. Crampton

Introduction

In 1938 two new terms entered the literature. Both were modern neologisms derived from Greek etymologies, and both were coined to describe geographic distributions. Both were proposed by senior and well-respected figures within their fields and both terms are still in popular usage today. Yet these terms indicate very different modes of thought concerning space, and ultimately questions of mapping, governance, and the biopolitics of race. In one case, the word "choropleth," which was coined by J.K. Wright (Wright 1938), President of the American Geographical Society (AGS), the term entered the geographical and cartographical literature immediately. It describes the most popular form of thematic mapping and GIS practiced today, in which geographically bounded regions are created. The second term is "cline," which was coined by Sir Julian Huxley (Huxley 1938) a British evolutionary biologist, first Director of UNESCO, and President of the British Eugenics Society, to refer to the gradual and continuous geographical variation of species, and more generally any continuous spatial variation. This term is well known in disciplines such as biology and anthropology, but has almost no presence in human geographic or cartography.

In this chapter I would like to respond to the invitation to rethink maps by examining the relationship between "mapping knowledges" and race in light of these two terms. How do maps frame our understanding of spatial distributions such as race, and how as a practice do they create and promote certain forms of knowledge and not others? In particular I examine types of maps including the choropleth, which treats space as a set of areally bounded units with discrete borders over which a property is extended: *res extensa* (Elden 2001). I contrast this approach with the anthropological and ecological use of "cline," which is used to understand spatial distributions such as human variation and race as continuous phenomena without inherent boundaries. I draw a contrast between these forms by reading them through the Platonic dialog the *Timaeus* and its discussion of being and becoming and apply this to recent discussions of ontology and ontogenesis.

There has been much interest over the last decade or so on boundaries, de-bordering, hybrid borders, and transboundary spaces in a globalizing world (Fall 2005; Newman 2006a, 2006b; Paasi 2002; Painter 2008; van Houtum *et al.* 2005). Yet relatively little attention has been paid to how mapping generates very specific territorial knowledges, or what Gunnar Olsson (1998, 2007) calls "cartographic reason." In this chapter therefore, I would like to try and connect up a critique of cartographic reason with this reappraisal of territory and boundaries. This chapter is also part of a larger critique of mapping knowledges involving the choropleth map (Crampton 2003, 2004; Sui and Holt 2008; Wright 1938) whose limitations, although long known within the technical literature, show no sign of causing a decline in its popularity.

Two terms: choropleth and cline

The different trajectories taken by the terms choropleth and cline are emblematic of the different kinds of spatial knowledges that they can and have produced. Here it is worth briefly examining how the terms arose and how they have been deployed.

The word "choropleth" was introduced by John ("Jack") K. Wright (1938). As Director of the American Geographical Society (1938–49) Wright was well respected in geography and he contributed widely to cartography since he was first appointed as Librarian in 1920 at the AGS. His position at the AGS afforded him the opportunity to know many leading geographers of the day. As I have discussed elsewhere (Crampton 2004) Wright was one of the first to formalize cartographic knowledge (e.g. see Wright 1944) by systematizing and connecting types of map with different representations of space—points, lines, and areas. This schema today underlies geographical information systems (GIS) and the vector spatial data models of points, lines, and polygons. David Lowenthal (1969: 598) called Wright's career "one of the most fruitful and illustrious in the history of American geography."

The word choropleth is usually translated as a bounded space (χώρα or *chôra*) over which a mass or throng (*plethos*) is extended. In this way the word is understood in terms of the well-known Cartesian *res extensa* (extension in space). As Wright himself put it:

> the term choropleth, which expresses the idea "quantity in area," is tentatively proposed. A choropleth is an areal symbol . . . [which] indicates densities as actually calculated for the areas that they represent. In the category of choropleth maps would be included maps on which the areas of differing densities are limited by the boundaries of administrative divisions and also maps on which the densities are differentiated within these boundaries.
>
> (Wright 1938: 14)

In other words the choropleth map is made from pre-existent bounded areas over which a value is extended ("quantity in area"). This single value is assigned to an enumeration unit and is derived by taking a space (enumeration unit) and counting up all the items in that space and then making an average to represent the variation. This average is then *extended* over the entire space. For example, we might map household income. In an enumeration unit there might be 2,000 households. We would take each household's income, aggregate it, and then divide by the total number of households:

$$\frac{\text{SUM } (i = N)}{N}$$

A choropleth map then uses predetermined but arbitrary boundaries of areally aggregated data extended over space. Figure 2.1 shows a typical choropleth map of racial prevalence based on U.S. Census Bureau data.

As I argue below, this understanding of the word *chôra*, which has been repeated in cartography textbooks for over half a century represents a radical forgetting of its original meaning. This earlier meaning, most famously used by Plato in his dialogue the *Timaeus*, is very different from the modern meaning of something extended over space as a bounded area.

It is significant therefore that the choropleth map is one of the most common thematic maps produced today (Slocum *et al.* 2008). Commonly used to

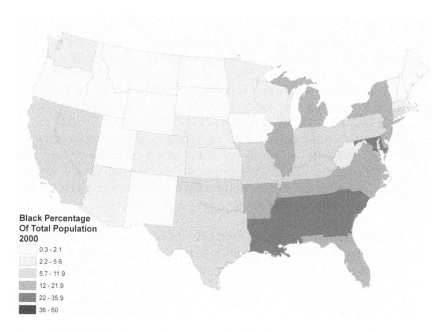

Figure 2.1 Typical choropleth map showing black percentage of total population. Source: author, 2000 U.S. census data.

represent statistical data, it is the dominant mapping type used by social scientists, the popular press, statisticians, and the government and electoral geographies. They are a default option in mapping software and a popular choice by students in cartography classes. A recent study found that about 60 percent of all maps published in leading public health journals published between 2000 and 2004 were comprised of choropleth maps (Martin 2005) despite their limitations for analysis of health distributions. Similarly, the principle behind choropleth maps of analysis by groups is commonly used in spatial profiling and risk assessment.

Choropleth maps are characterized by a number of well-known limitations, including the ecological fallacy and the modifiable areal unit problem (i.e. the mapped values are dependent on how the boundaries are arranged). The ecological fallacy is a common obstacle to geographic analyses and states that it is incorrect to infer individual level data from areal units. For example, using Figure 2.1 to infer specific rates for localities within states would be fallacious. Non-uniform distributions are particularly hard to interpret (Tate *et al.* 2008). For these reasons the choropleth map is considered a weak form of spatial analysis.

Why are these maps nevertheless so popular? Sui and Holt (2008: 4) suggest that there has been a "flattened learning curve" with regard to spatial analysis in public health and disease mapping and a consequent reduction in "cartodiversity"—analogous to biodiversity. There is some evidence for this. Overviews of the historical record of disease mapping reveal that it is far richer than it is today, with a range of available map types including dot maps, dasymetric maps, isarithmic maps, and cartograms (Koch 2005), not to mention now "extinct" forms such as the "isontic," "chorisopleth" or "chorogram" (Wright 1944) or "three-dimensional thermo-isopleth" (Harold-Smith 1929). Why has there been such a loss of "cartodiversity" and what does it portend for our understanding of spatial distributions and hence for political governance?

The second term is "cline," which derives from the Greek word "*klino*" or slope, slant, incline and was proposed by Sir Julian Huxley to refer to "a gradation in measurable characters" (Huxley 1938: 219). Today the word is used to refer to gradual changes that occur, such as gradual or continuous changes in genetic characteristics over space. Huxley originally envisaged a suite of words depending on the specific application: ecocline, genocline, geocline, chronocline, and ontocline, but these have not survived in general usage. Cline therefore means that we should understand biological variation as continuous and gradual, rather than discretely. In this context Huxley rejected the notion of race: humans vary, but continuously and cannot be discretely categorized (especially spatially). Figure 2.2 is an example of a blood group cline from an anthropology textbook.

Huxley, a Fellow of the Royal Society, was the Secretary of the London Zoological Society at this time and would later become the first Director of UNESCO. One of Britain's leading evolutionary biologists, he was knighted

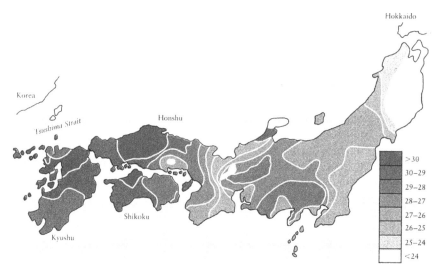

Figure 2.2 Clines in the percentage of blood group A in Japan. Source: Lewontin
1995: 116. (Copyright 1995 by Scientific American Library. Reprinted
by permission of Henry Holt and Company.)

in 1958. He came from a notable family—his brother was Aldous Huxley,
author of *Brave New World* and his grandfather was T.H. Huxley, the friend
and supporter of Charles Darwin. Huxley introduced the term cline to
counteract what he saw as defects in the understanding of geographical
distributions:

> It is in no way intended that specification by clines should replace any
> of the current taxonomic methods. It would constitute a supplementary
> method which, it is suggested, would correct certain *defects inherent in
> that of naming areal groups*, notably in stressing continuity and regularity
> of variation as against *mere distinctness* of groups.
>
> (Huxley 1938: 219, emphasis added)

Thus, in terms of race:

> Clines are variations in the intensity of expressions of known hereditary
> traits over wide geographic regions. Skin colour represents a prime
> example of such a cline, since its gradations are continuous and can be
> plotted on a map showing its correspondence to latitude and temperature
> variations.
>
> (Smedley 1999: 315–16)

Other common examples includes clines of blood type or of prevalence of
a particular gene over space. The understanding here is not of areas, as with

the choropleth map, but of graded variation in intensity. As applied to human variation, this understanding came to prominence within anthropology nearly 50 years ago when Fredrik Barth problematized the boundary in anthropology (Barth 1969). By 1962 it was possible to summarize by stating "there are no races, there are only clines" (Livingstone 1962: 279; see also Gould 1977).

These two words, then, refer to very different pictures or understanding of spatial distributions. On the one hand, the choropleth is a map that partitions space into bounded areal units such as counties, census tracts, or zip codes. The cline treats spatial variation as continuous or gradual change. I argue that these different terms indicate very different modes of thought in the treatment and understanding of space and that the choropleth is an inadequate understanding with important implications for the production of racial categories.

It should be clear that I am not suggesting that these terms either dictated or introduced new ways of thinking. Rather, the terms were called forth to name ways of thinking. Certainly both areally bounded spaces and clinal variation can be traced back much further. Spatial aggregations are a form of categorizing that is often useful when confronted with variation or multivariate data. Clines are akin to statistical distributions such as the normal distribution; that is, continuous variation. Both can have their place, but categories suffer from the problem that they can become naturalized and immutable—as the history of racial categories has repeatedly demonstrated. When the term is applied to space a similar essentializing may occur and certain spaces begin to acquire a transcendent and sometimes privileged identity—what Hard (2001) calls "landscape racism."

By explicitly contrasting the two terms we can examine the ways in which people have "thought out space" (Foucault 1984: 244). In the remainder of the chapter I will trace out the varying ways in which space is thought out or calculated and its implications for a rethinking of mapping. In the next section I will pick up the suggestion made earlier that the modern translation of *chôra* as the Cartesian space of extension is an inferior understanding by turning to Plato's well-known dialog the *Timaeus* where the term is treated in some detail.

Chôra in Plato's *Timaeus*

Plato's *Timaeus* is a cosmogony or origins story. It is in the form of a dialog between Socrates and Critias, Hermocrates and Timaeus himself. It is a sequel of sorts to *The Republic* in which Plato tries to explain not only the physical origins of the universe but also its philosophic and metaphysical principles. Although it is a difficult and sometimes obscure dialog it also contains some famous moments, such as the Atlantis myth and the origins of the universe at the hands of a benign craftsman. In modern times it has provided a wealth of discussion on spatiality, mind and body, and philosophy, particularly in the writings of Julia Kristeva and Jacques Derrida (Grosz 1995).

Plato discusses the structure of matter, the evolution of the human form (said to have originated as a spherical head that rolled around "unable to get over or out of its many heights and hollows" before growing limbs), astronomy, and metaphysics. Among the latter, Plato distinguishes between "being" and "becoming" by elaborating on his famous story of shadows on the cave wall in the *Republic*. There Plato proposed that there is an ideal form of the universe that can be conceived of by us but never apprehended with the senses, and a perceptible universe that is a copy and never "fully real."

In *Timaeus* Plato considers the question this way:

> We must in my opinion begin by distinguishing between that which always is and never becomes from that which is always becoming but never is.
>
> (*Tim.* 27d)[1]

This distinction is central to recent debates in GIS and mapping opposing being (ontological entity definitions—"ontologies") and becoming (processes or mapping practices). Consider, for example, the following two approaches. From the computer science perspective of formal entities:

> an ontology is a formal universe in which each entity is precisely defined and its relationship with every other entity in the specific categorical or computing realm is also predetermined. Ontologies in this context are the range of what is possible. They can be thought of as simply a classification system or a data dictionary.
>
> (Schuurman 2009)

By contrast consider Kitchin and Dodge's (2007: 335) argument emphasizing mapping *practices* and becoming, rather than being:

> we are outlining what we believe is a significant conceptual shift in how to think about maps and cartography (and, by implication, what are commonly understood as other representational outputs and endeavours); that is a shift from ontology (how things are) to ontogenesis (how things become)—from (secure) representation to (unfolding) practice.

These two orientations are not new in concerns but hark back to Plato's distinction between being and becoming. Although Kitchin and Dodge's specific formulation for mapping is original, I would argue that it draws from a long critical tradition in cartography that is not so much concerned with the form or "look" of the map as famously outlined in Arthur Robinson's influential post-war text (Robinson 1952) but with power–knowledge relations. This "critical" tradition although often associated with the work of Brian Harley and Denis Wood beginning in the 1980s can in fact be found in much earlier work by authors such as J.K. Wright (1942) and Mark Jefferson

(1909) as I have argued elsewhere (Crampton 2009; Crampton and Krygier 2006). Nevertheless, in my view Kitchin and Dodge's discussion pushes very directly at the question of knowledge and mapping.

The word *chôra* occurs nearly a dozen times in Plato's *Timaeus*. The Greeks used several words for spatial terms, including *topos* or "place" in addition to *chôra*. The standard translation is "space [territory] in which a thing is" with an implication of the proper or fitting place; one's post, station, position (Liddell and Scott 1990/1871: 793). In Greek there was also a word *chorophilia* meaning "love of a place," to haunt or frequent a place [χωροφιλέω]. Χώρα might also mean the country, as distinct from the city; *chôra* rather than *polis*. In one variation, this could occur in a slightly demeaning manner as "rustic" or "boor," or a kind of country bumpkin [χωρίτης] out in the sticks.

This sense of an exteriority, of an outside or marginal place is exemplified in Plato's first mention of *chôra* in *Timaeus*. Socrates discusses the arrangements in his ideal society for children who do not make the grade (*Tim.* 19a): they would be put into a place [*chôra*] *outside* the citadel proper. As one commentator notes "from the outset what would be said in this word is posed at the margin of what can be fabricated, marking the limit of controlled production" (Sallis 1999: 19). Sallis refers here to fabrication (*poesis*) and production as the result of techniques or arts (*techne*) that bring something new into being; that is becoming or ontogenesis. Foucault's "technologies of the self" arise from this sense of the production using techniques and methods in a practice, or in Greek an *askêsis* (McGushin 2007). *Chôra* therefore is not produced in an ordinary manner (e.g. as becoming); it is precisely a difficult and problematic concept.

Sallis, drawing on the Derridean deconstructive tradition, denies that *chôra* is the "isotropic space of post-Cartesian physics . . . nor is it even empty space" (Sallis 1999: 115). In the dialog Plato explains it by introducing a third term in addition to and different from being and becoming. This is *chôra* as the famous "receptacle" for becoming:

> We must start our new description of the universe by making a fuller subdivision than we did before; we then distinguished two forms [kinds *eide*] of reality—we must now add a third kind [*triton allo genos*] . . . we postulated on the one hand an intelligible and unchanging model [being] and on the other a visible and changing copy of it [becoming]. We did not distinguish a third form . . . a form that is difficult and obscure. What must we suppose its powers and nature to be? In general terms, it is a receptacle and, as it were, the nurse of all becoming and change.
>
> (*Tim.* 48e–49a)

This third form or kind is not a new kind of being or a new kind of becoming because those have already been identified. To explain, Plato immediately moves to a discussion of fire, which is ever-changing and without fixed properties, yet seems to have a "fiery" character. Attempting to grasp this "difficult and obscure" notion he offers more ways of thinking about it; it is like a malleable metal that can be made into many forms, or like the base alcohol of a perfume that can be given different aromatic notes (50b–c). It is concluded that there are three kinds: 1) becoming; 2) that "wherein" it becomes; and 3) the source "wherefrom" becoming is copied and produced (Bury translation, 50c–d). The "that wherein it becomes" is the *chôra*, the space where being is birthed (*genesis*, 49d) or, in Kitchin and Dodge's (2007) terms "ontogenesis."

Thus we can interpret this in the sense that *truth is produced by the very act of mapping*. What I find interesting about this is that it is such a distinctly political project. The map does not record static, pre-existent beings (the "confession of the landscape" as it were) but is itself the act of making truth. To the extent that we forget this struggle in the choropleth map (by treating the map as an object, an object for calculation, and space as extension) we have moved away from our proper engagement with the politics of the map.

Following Heidegger, Elden (2001) suggests that an original Greek understanding of space as "situated place" was later eclipsed by a Cartesian understanding of space as "bodies extended in space" (Elden 2001: 32) in the old sense of space as a container. Thus, argues Heidegger:

> the Greeks had no word for "space" [as *res extensa*]. This is no accident; for they experienced the spatial on the basis not of extension but of place (*topos*); they experienced it as *chôra*, which signifies neither place nor space but that which is occupied by what stands there.
>
> (Heidegger 1959: 66)

In sum, then, the *chôra* is neither being nor becoming. It is not a "space" that is fixed and changeless, but nor is it pure becoming and not being. It is productive and generative space (the "nurse of becoming"). The modern word "choropleth" and the meaning of space as just an area is very remote from these original meanings and one cannot but help think that something richer has been lost in the process.

Mapping and identity

Let me now return more explicitly to the question I posed at the beginning: how do maps frame or produce political knowledges? In particular I would like to examine how the choropleth map as *res extensa* compares with the clinal map in the production of identity (especially race) of populations (rather than of a specific individual).

Efforts to spatially characterize human populations extend at least as far back as the classical Greek and Roman times. In his *Histories* (written around 430 BC) Herodotus did not hesitate to describe the various peoples in far-flung parts of the "oikumene" (inhabited world) however strange (he dutifully records both cannibals and werewolf men, *Hdt.* 4.18; 4.105, though he disbelieves the latter). His geographical locations were fairly good, if general. There was no attempt to make precise boundaries. Though peoples were sometimes described as occupying particular locales, these were not political borders (indeed Greece itself was not a country in the modern sense, rather it was a series of often warring nation-states). Other works such as Pliny the Elder's 36-volume *Natural History* (first century AD) also provided descriptions of a panoply of human variation.

After classical times there have been numerous attempts to delineate population groups with geographical spaces or areas. The mappa mundi (world map) of Isodore of Seville (died 636) divided the world into three partitions, according to the Biblical tradition in Genesis (Chapter 10) that the three sons of Noah (Shem, Japheth, and Ham) peopled the earth (Wallis and Robinson 1987). Thus the world was populated by three major groupings of peoples (Asians, Europeans, and Africans). Later medieval mappae mundi such as the Hereford map (*c.* 1300) supplemented this scheme by drawing on travellers' tales, Biblical information and Herodotus to put illustrations of strange peoples on their maps.

The increasing sense in the West of an "us" and "them" can be traced back to the encounters produced following the great age of discovery after the fifteenth century. According to the anthropologist Jonathan Marks this sense of the "other" gradually expanded from a quite local one, perhaps of the village in the next valley, to continent-wide designations during the nineteenth century, that is to the idea of geographical races (Marks 1995, 2006). As Winlow (2009) has discussed, the establishment of evolutionary theories in the nineteenth century served to redouble efforts on mapping human racial types as part of a whole concern with human characteristics, population density, migration, longevity, and especially language and religion. These latter two could and did often act as surrogates for "race."

Early clinal maps are known from at least 1701, when Edmund Halley produced a thematic map using isolines to show magnetic declination across the globe, and a century or so later Alexander von Humboldt presented a more refined technique of "isotherms" (Robinson and Wallis 1967). We should also note that a concept known as the "isocline," or line of equal slope, was developed at this time (early nineteenth century), which is still today used in population dynamics. However, these were not applied to human distributions until the nineteenth century.

By the mid nineteenth century multiple forms of mapping were in use to understand human populations, including surface (or clinal) maps, the choropleth, and the dasymetric, which was named (but not originated) in 1923 as a reaction against the choropleth map (McCleary 1969). The subject matter of these maps was initially population density, but during the nineteenth century more refined understandings of human population groups were developed. Maps were made of mortality, education, crime, longevity, language, religion, birth and death rates, age of first marriage, and so on.

These subjects were of concern as "moral statistics," or how best the modern state should be governed (Hacking 1982).

By the nineteenth century theories of human origins and the tools of statistical distributions could be applied to construct ethnographic maps, often based on language (such Ami Boué's 1847 map of Turkey in Europe; Boué was an Austrian geologist who went to Edinburgh in the aftermath of the Scottish Enlightenment and studied under Robert Jameson) but also explicitly on race, such as Gustav Kombst's 1846 map in A.K. Johnston's *The Physical Atlas*.[2] In the U.S.A. the first statistical atlas of the census contains many of these new race-based maps (Hannah 2000; Walker 1874). Hannah has argued that the particular concerns with immigration of Francis Amasa Walker, the Superintendent of the U.S. Census Bureau for the 1870 census, were lent a powerful scientific impetus by the race-based maps he produced in the *Atlas*. Writing in the *Atlantic Monthly* in 1896 Walker argued that immigration restrictions should apply not only to hundreds or even thousands of people but to hundreds of thousands, and not because they were deaf, dumb, blind, or criminal, but simply because they would subject America to "degradation through the tumultuous access of vast throngs of ignorant and brutalized peasantry from the countries of eastern and southern Europe" (Walker 1896: 823; see also Sluga 2005). In other words the quality of the American population would be reduced by what was at the time the commonly identified threat of south-eastern Europeans (Italians, Slavs, Greeks, Hungarians, and so on).

Today, physical borders may similarly act to produce identity, for example, a national border may serve to construct a "biometric identity" when they are crossed or approached (Amoore 2006; Häkli 2007). In this case, identity construction depends upon risk profiling by placing individuals into certain groups and performing a calculation about that individual's riskiness as a consequence of which groups he or she may be a member of (Amoore 2006; Bell 2006). The "risk" of such risk assessments is that they can produce far more false positives than true positives, thus draining labour power and resources (Crampton 2007). There has now developed a growing literature on this aspect of identity construction, especially looking at it from the perspectives of biopolitics (Alatout 2006; Foucault 2008; Legg 2005; Lemke 2001).

Unfortunately as yet we still do not have a comprehensive history of mapping and the terms in which it frames (racial) identity. Thanks to early and important work by Robinson (Robinson 1971, 1982) we do know that the isarythmic or surface maps were used long before the choropleth map, perhaps as early as the sixteenth century (by contrast the first known choropleth was produced in 1826). MacEachren (1979) discusses the history of thematic mapping using point, line, and area-based map types drawing largely on Robinson (and ultimately Wright (1944) though he is not cited). And there is well-known work in GIS by Langford, Unwin, Martin, Maguire, and colleagues (Langford and Unwin 1994; Martin *et al.* 2000; Tate *et al.* 2008) that explores surface modelling methods of population estimation. The latter writers have explicitly rejected areal aggregation approaches such as the choropleth, and instead pioneered a number of innovative techniques for analyzing continuous distributions and the use of remote sensing data to estimate population density (e.g. see Langford *et al.* 1991). Cognate work of organizations such as Oak Ridge National Laboratories and their high resolution (three arc seconds or ~90 meter) "Landscan" dataset is also extremely valuable in mapping human populations with remotely sensed imagery (Bhaduri *et al.* 2007).

Despite this gap in the literature, I think we can adduce a few clues that will help us understand how and why mapping has been deployed in the production of population identity. The earlier development of the surface or isarythmic map should not distract us. Clinal maps of this type were readily applied to the natural world and to the rather later arriving practice of recording statistics about human populations occupying territory (for example, the establishment of the modern U.S. decennial census in 1790 and the UK census in 1801).

However, this observation only serves to force us back to the question of why and how concern for territorial occupation by certain human populations originated. Here I shall draw on Foucault's discussion of the origins of the modern state as a scheme of biopolitics or biopolitical governance. This is not meant to imply that Foucault has exclusive access to useful analysis. Bruno Latour's work on knowledge circulation and assemblages, or how things get linked into a collective (including non-human things) has probably been more influential in geography and cartography (on the latter, see especially Turnbull 2003). Latour's account of space in Actor-Network Theory (ANT) as networks, assemblages, and flows is readily applicable. In this light it would be interesting to have a detailed account of the ways these two authors complement each other (for example, both talk about techniques and technologies).

Foucault's new Collège de France lectures series do allow us to explore Foucault's work on space and afford the opportunity to link this back to mapping practices (Crampton and Elden 2007). Three lecture courses in particular are germane here (Foucault 2003, 2007, 2008). Foucault delivered these lectures during the late 1970s and these new books represent edited transcriptions of the delivered remarks (Foucault famously left a letter indicating "no posthumous publications" but his heirs and family accept that these lectures were public). After 1976 (1978 in English) when the first volume of the *History of Sexuality* (Foucault 1978) was published Foucault entered a period in which he was intensely interested in the political production of knowledge. Although the 1976 book is famous for its discussions on sexuality, it is in the last section, entitled "Right of Death and Power Over Life" (which Foucault claimed in an interview was always overlooked), that we can see this theme emerging. This section in fact has little to do with sexuality and much more to do with "political" knowledge, and the way that historically, certain forms of knowledge have been promoted or deployed. It was this theme that occupied much of the rest of his work in the 1970s and which lies behind the increasing interest in the biopolitical in geography (though not cartography or GIS as yet).

Foucault's argument was fundamentally straightforward. He claimed that the modern state, as it moved away from pure sovereignty, adopted a more "governmental" approach. Whereas with sovereignty the issue was often one of control and discipline of individuals or a "micro-physics of power" (Foucault 1977: 26) at some point many modern states realized (explicitly or implicitly) that it was neither necessary nor possible to regulate individuals. Instead, the modern state (which Foucault dates from approximately the seventeenth century onwards in this context) began to adopt a "biopolitical" approach, which from about 1978 onwards he started to call governmentality or the art of government (Foucault 1978, 1991; Lemke 2001).

Governmentality has received copious attention from scholars over the last two decades and it is not my intention to discuss it directly here except insofar as it relates to the production of cartographic identity (for summaries of these Foucault lectures see Elden 2007). Foucault's notion of the biopolitical highlighted a different form of governance alongside that of discipline; namely the governance not of individuals but of populations. For Foucault an explanation of power relations as comprising only sovereignty began to look inadequate. There were far more relations of power-knowledge than just the top-down ones emanating from the monarch or Machiavellian prince, there were rather "multiplicity of subjects, or ... the multiplicity of a people" (Foucault 2007: 11) (echoes of Latour here). "Population," claimed Foucault (2007: 11), "is undoubtedly an idea and a reality that is absolutely modern in relation to the functioning of political power."

The multiplicity involved here is therefore not just one of many relations of power-knowledge, but also the recognition that there are a whole series of "mobile elements" that have to be managed and which can only be known by a distribution of probabilities. If sovereignty was concerned with the seat of government in a territory and discipline with the structure and hierarchy of a territory, then now we are dealing with a *milieu* or "the medium of an action and the element in which it circulates" (Foucault 2007: 21).

This notion in Foucault of milieu and its relation to governance and biopolitics has rather surprisingly been overlooked by geographers. Of course, the term has been influential before, notably in the French Annales school (Buttimer 1971). But here it has a different relevance because Foucault draws on its usage by the Count de Buffon, the French naturalist who led a movement opposed to Linnaean classification in the eighteenth century. It will be recalled that the notion of the "species" is an eighteenth-century one and that Linnaeus placed it in a set of nested categories (species—genus—order—class). Instead of this classificatory scheme, argued Buffon, there were only species, but within species the environment (milieu) could cause population differences (but not new species).

For Linnaeus, therefore, "race" is a subcategory of the species. In the 1740 edition of his work Linnaeus posited four geographical subdivisions of humans: white Europeans, red Americans, yellow Asians, and black Africans. However, this scheme left Linnaeus with a mixed bag of other human forms (such as the small Alpines peoples and the "flatheads" of Canada) that did not fit with any distinct geography. Thus, as Marks (1995: 50) observes, his scheme was nothing more than a "socio-cultural" scheme masquerading as a biogeographical one (especially as he used quite derogatory terms for the non-whites in his scheme).

Where Linnaeus had classification as his goal (isolation of common elements), Buffon had *diversity* as his (explaining variation). Unfortunately for anthropology, says Marks, between 1758 (Linnaeus' tenth edition) to the 1960s (when it was overthrown) physical anthropology followed Linnaeus in searching for races and their nature:

> It is one of the blindest alleys in the history of modern science. The question ignores the cultural aspect of how the human species is carved up; it ignores the geographically gradual nature of biological diversity within the human species [i.e. clines] and it has a strong anti-historical component in its assumption that there was once a time when huge numbers of people, distributed over broad masses of land, were biologically fairly homogenous within their group and different from the (relatively few) other groups.
>
> (Marks 1995: 52)

In other words it is clines and not choropleths. The choropleth map has contributed to this blind alley, whereas understanding the range of biodiversity in human populations allows us to see how they vary gradually with environment, migration, and genetic drift. These are the mechanisms of variation. Populations as well as individuals reproduce themselves, although like individuals not always identically due to mutation, and the introduction of new genetic material through intermarriage and migration (gene flow). These changes in the gene pool are then passed on through natural selection, but gene pools can also be affected by non-adaptive fluctuations known as genetic drift that are not caused by the environment. Or, to put this in a way that harks back to the discussion above, it is the difference between being and becoming. And the milieu is the *chora* in which becoming is nursed and produced.

By setting up categories of opposition (such as races) and other identities rather than a graded geo-biodiversity, we are partaking in a rather modern discourse of partisanship. Rather than a unitary or universal perspective it is a discourse of opposition. Foucault goes so far as to say it is a discourse of war that underlies apparent "peace" and he inverts Clausewitz's famous dictum to read "politics is war carried on by other means" (Foucault 2003). In the development of biopower, suggests Foucault (2003: 61), we need to understand that it was predicated on a discourse of races waged by different partisans *within* the society as a whole and not as one society against another:

> It is the splitting of a single race into a superrace and a subrace . . . It will become the discourse of a battle that has to be waged not between races, but by a race that is entitled to define the norm, and against those who deviate from that norm, against those who pose a threat to the biological heritage.

If Foucault is right the rise of the biopolitical, with its emphasis on births, deaths, marriages, degeneracy, and so on can be read through as a concern with the efforts to establish and protect a normalizing society. And mapping, to the extent that it partook in this discourse, played along with that effort. Thus the choropleth is complicit in marking out these areas or regions of concern. For Foucault (2003: 255–6) then, racism:

> is a way of introducing a break into the domain of life that is under power's control . . . the appearance within the biological continuum of the human race of races, the distinction among races, the hierarchy of races, the fact that certain races are described as good and that others, in contrast, are described as inferior.

Foucault recognizes that to cut into the "biological continuum" is absolutely necessary in a normalizing biopolitical society. Far from the continuum such as those mapped by clines, we obtain groups based on difference.

As an illustration of this point we can briefly consider an important work carried out just prior to World War I by a member of the American Geographical Society (AGS), Leon Dominian (1917). Dominian's book was an attempt to show the relation between language and nationality with a view to settling political boundaries. An "ill-adjusted boundary is a hatching oven for war. A scientific boundary . . . prepares the way for permanent goodwill between peoples" he begins (1917: vii). Dominian pays particular attention here to the "Eastern Question" (the problems posed by the decline of the Ottoman Empire) and how geographical knowledge could provide an acceptable settlement. This book was part of an increasing shift from a geopolitics dependent solely or mainly on "natural borders" (ridge lines, rivers, watersheds) as providing defensible boundaries, to one that increasingly incorporated "population borders" whether using race, language, religion, economic trade, and so on.

Dominian used both, arguing that borders began in nature but were elaborated by humans, and that natural borders then fade out as "the result of man's progress . . . [by] the removal of natural obstacles; the conquest of distance by speed" (1917: 327). Dominian himself highlighted economic development, and it is the more remarkable then that his book should have an Introduction by Madison Grant, author of perhaps one of the most notorious racist tracts of the early twentieth century (Grant 1932, first edition published in 1916). In fact Grant was an AGS Councillor for several decades and his book first appeared in its journal the *Geographical Review* (Grant 1916). For Grant, race was a meaningful biological (phenotypic) variable: "race taken in its modern scientific meaning [is] the actual physical character of man" (Grant 1917: xv) and "it is entirely distinct from either nationality or language . . . race lies at the base of all manifestation of modern society" (Grant 1932: xxi). Race, for Grant, was a substrate written into human biology. It is neither a linguistic nor a political group (Grant even observes this in Dominian's book, warning him of seeing race in his linguistic groups). Even achievements made by non-whites were the result of "mimicry" of whites imposed from without by social pressure of the "slaver's lash":

> Whenever the incentive to imitate the dominant race is removed the Negro or, for that matter, the Indian, reverts shortly to his ancestral grade of culture. In other words, it is the individual and not the race that is affected by religion, education and example. Negroes have demonstrated throughout recorded time that they are a stationary species and that they do not possess the potentiality of progress or initiative from within.
>
> (Grant 1932: 77)

Race then is inherent and fixed, biological, it is the essence of being. For Grant there were three major races in Europe (Nordic, Mediterranean, and Alpine). Outside Europe the major races were "Negroid" and "Mongoloid" (Grant 1932: 32). Some countries were more affected by archaic or "Neolithic" traits; Britain for example, although admirably Nordic (blond, blue-eyed, flowing hair) in general, sometimes yielded evidence of this less-developed trait. Who can fail to observe, says Grant (1932: 29) "on the streets of London the contrast between the Piccadilly gentleman of Nordic race and the cockney costermonger of the old Neolithic type?" Women of all races also exhibit "the older, more generalized and primitive traits of the past of the race" (Grant 1932: 27).

Grant vociferously denied that this substrate could be molded by the environment, a central tenet as we have seen of the explanation for genetic changes in a population. There is "a widespread and fatuous belief in the power of the environment, as well as of education and opportunity to alter hereditary, which arises from the dogma of the brotherhood of man" (Grant 1932: 16). Grant sarcastically makes fun of emerging anthropological findings that even head shape could change among immigrant groups, a finding first discovered by Franz Boas and now generally accepted in anthropology.

As was written in his own book's Preface:

> conservation of that race which has given us the true spirit of Americanism . . . if I were asked: what is the greatest danger which threatens America today? I would certainly reply: the gradual dying out among our people

of those hereditary traits through which the principles of our religious, political and social foundations were laid down and their insidious replacement by traits of less noble character.

(Grant 1932: ix)

In these terms we see the exact principle of biopower identified by Foucault: the concern for the within, not so much the concern for the without. Where stands the stock in terms of the master and the "less noble character"—the "superrace" and the "subrace"? That is the meaning of race and racism. The writer was Henry Fairfield Osborn, a leading evolutionary biologist of the American Museum of Natural History.

Dominian's book itself not surprisingly deploys many maps that attempt to delineate the geographical extent of various languages (see Figure 2.3). His task here may seem quite a daunting one given the fact that the predominant language of a region may not be the only language of a region, and that dialects within a language add an additional complication. However, Dominian, although perhaps not in a position to point out differences with his employer's Councillor, does not take the same approach as Grant. For Dominian, what is important is not so much the inevitability of the racial substrate, as the effects of the environment, of economics and of human development; the non-biological. Although he believes heredity is important he takes a much wider approach in understanding human variation, perhaps what we would today call one of "nature-culture" (Goodman *et al.* 2003).

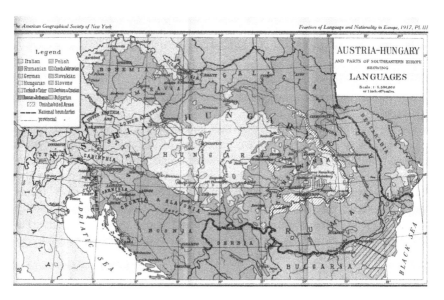

Figure 2.3 Language distribution in the Austro-Hungarian region. Source: Dominian 1917.

Nevertheless this was still an exercise in finding bounded regions with a view to establishing "scientific" political borders.

Cartographic calculation of race and ethnicity

During the 1960s anthropologists began to grapple with the effects of borders and boundaries and the question of ethnicity (another term promoted by Huxley in favor of "race"). This work tended to favor flow and change more than stability and fixity. Even if there were ethnic groups, these were cultural and not biological and even if they have boundaries (social or territorial) these boundaries are not impermeable barriers but crossable, subject to erasure and flows, disappearing and coming into being again (Barth 1969). By the early 1970s anthropologists could also draw on new research in genetics which showed that about 85 percent of the total genetic variation in humans comes from individual's differences within groups and only about 15 percent to differences between ethnic groups or populations (Lewontin 1972). For example, there is more genetic difference between one black person and another than between black and white people as a whole. Anthropological "statements on race" such as those adopted by the American Association of Physical Anthropologists (AAPA) in 1996 and the American Anthropological Association in 1998 both emphasize that race has no biological or geographical basis: "humanity cannot be classified into discrete geographic categories with absolute boundaries." They use both genetic aspects of heredity and the environment to explain human variation, which occur as gradual changes in clines. To my knowledge the leading Anglo-American geographical organizations have never issued statements on race despite their longstanding interest in race, racism and the widespread usage of race-based data.

Furthermore, with the rapid expansion and adoption of GIS, GI scientists have developed a "Body of Knowledge" as the basis for GIS expertise which includes ethical and societal issues (DiBiase *et al.* 2006). It is also used as the basis for awarding nationally accredited certificates in GIScience, the "GISP," from the GIS Certification Institute. The Body of Knowledge effort was recently also recommended by a National Research Council committee examining the importance of GIS in meeting "national needs" as it "provides the basis for determining the eligibility of education achievement claims for GIS certification" (National Academy of Sciences 2006: 57). Neither the Body of Knowledge nor the NRC committee address the questions of identity, race, the ethics of using race-based data, or power. The Body of Knowledge is explicitly a fact-based document rather than a process-oriented one emphasizing critical thought.

In this light it would be useful for efforts such as the Body of Knowledge to grapple with the cartographic production of knowledge and the difference it makes to think clinally. Have choroplethic methods merely endured because they are easy to conceive and to produce in modern day GIS? If so, a different type of GIS software would appear to solve the problem (or at least offer a wider "cartodiversity"). The slight resurgence in dasymetric mapping (another old form that has been used only rarely in modern times) might offer an instructive example (Eicher and Brewer 2001). On the other hand it may be more than just a matter of tools and also a matter of the political ends to which knowledges are put. Thus we might argue that a political rationality of government and populations that came to fruition in the early twentieth century is no longer adequate. But of course this is harder to dislodge when it appears to yield mechanisms for identifying "risky" populations.

Contributions to our understanding of identity and its relationship to space are nevertheless not difficult to find. Making territories is an exercise in power, boundaries are now understood as hybrid (Fall 2005), constructed, political, and heterogeneous. Mapping as a form of reason is critically important here. A recent paper reinvigorates Gregory's (1994) phrase of a "cartographic anxiety" concerning cartographic reason in the construction of regions and territory: "not only in the drawing of maps themselves, but also to geographical knowledge and other forms of knowledge more generally. In other words, a cartographic impulse may be at work whether an actual map is produced or not" (Painter 2008: 346). In this chapter therefore I have looked at both the actual drawing of maps themselves but also the cartographic reason that they help inform.

Conclusion

The *Timaeus* is a rich and frustrating dialog that I have puzzled over for many years. In fact nearly seven years ago I wrote a much earlier version of this chapter but was so dissatisfied with it that I put it away in a drawer. (The current chapter is 99 per cent original.) One of the reasons for this was the amazing disparity between what passes for our understanding of "chora" as space to be filled up and Plato's elusive and suggestive idea of it as the "nurse" of becoming. Two developments have helped me bring out something useful: my increasing dissatisfaction with the rise of computer science approaches to "ontology," which treat being as fixed and changeless, and Kitchin and Dodge's (2007) attempts to rethink mapping as ontogenesis. Although we approach these issues from somewhat different angles I see them as being complementary rather than contradictory.

If we read "chora" as the space in which becoming takes place, rather than its current meaning as *res extensa*, then obviously the question arises of what it is that becomes in space or milieu. The tradition of creating identity through a presumed occupation of a common homogenous space (the choropleth) whether it draws from language, ethnicity, or race is I think a desultory one. In this regard I have tried to think of identity not as a "group" but as a gradation—a cline, to use Julian Huxley's term. I think this emphasizes our commonality without putting impermeable and arbitrary barriers or borders around groups (cutting into the "biological continuum" in Foucault's words), but still allows us to explore differences. If clines are still unusual in human or political geography, they are well known and accepted in the study of human variation. In contributing to this rethinking of mapping then I suggest we use clines to explore human identity, rather than bounded areas or groups.

I believe this is best achieved through an interdisciplinary approach, in this case anthropology (and the concept of clines) and geography (and the concept of milieu). Nash (2003: 638) says that "as with 'race'":

> this [disciplinary identity] is an issue of boundaries and names for entities, objects and classes of people and things, that both makes communication possible and powerfully naturalizes differences and divisions through those names and categories.

The introduction of the term "cline" by Huxley in 1938 has not made race obsolete (nor, since racism is predicated on accepting that there is meaning to race, has it eliminated racism; while we have race we will have racism). But it has given a powerful array of tools for framing discourses of human variation as graded "geo-biodiversity" that has informed anthropological debate both within and outside the discipline. For instance, it has helped the discipline through its professional organization take a stance on the use of race-based data (e.g. from the census); a stance that is notoriously lacking in Anglo-American geography. Many anthropologists speak out publicly about the reality of race. (In one recent case following an op-ed by an evolutionary biologist in the *New York Times* seeking to resurrect biological race, the Social Science Research Council commissioned a series of responses from leading scholars and posted them on its site.)

And it is crucial to keep doing so. Although we might think that biological understandings of race such as that put forward by Madison Grant of the AGS have long since faded, writers are noting that it is making a rather unwelcome re-appearance as "genomic race" and race-based medicine (Duster 2005). We have already seen the first U.S. Food and Drug Administration (FDA) approved race-based drug, BiDil, marketed toward African-Americans. Geographers have long employed mapping to understand health distributions (Koch 2005) and GIS is increasingly used by public health professionals (including the U.S. Centers for Disease Control), a reliance on the choropleth will only help feed those categories. If racial understandings of disease are a "fallacy" in the words of one recent writer (Graves 2001) then the production of knowledge through mappings—the cartographic calculation of space— will continue to be a critical process.

Notes

1   I have used both the 1965 translation by Desmond Lee (revised 1977) (Plato 1977) and the 1929 Loeb edition by R.G. Bury (Plato 1929).

2   Kombst's map is reproduced on the David Rumsey website, www.davidrumsey. com.

References

Alatout, S. (2006) 'Towards a bio-territorial conception of power: Territory, population, and environmental narratives in Palestine and Israel', *Political Geography*, 25(6): 601–21.

Amoore, L. (2006) 'Biometric borders: Governing mobilities in the war on terror', *Political Geography*, 25(3): 336–51.

Barth, F. (1969) *Ethnic Groups and Boundaries. The Social Organization of Culture Difference*, Boston, MA: Little, Brown.

Bell, C. (2006) 'Surveillance strategies and populations at risk: Biopolitical governance in Canada's national security policy', *Security Dialogue*, 37(2): 147–65.

Bhaduri, B., Bright, E., Coleman, P. and Urban, M.L. (2007) 'Landscan USA: A high-resolution geospatial and temporal modeling approach for population distribution and dynamics', *GeoJournal*, 69(1–2): 103–17.

Buttimer, A. (1971) *Society and Milieu in the French Geographic Tradition*, Monograph Series of the Association of American Geographers, No. 6, Chicago, IL: Rand McNally for the AAG.

Crampton, J.W. (2003) 'Are choropleth maps good for geography?', *GeoWorld*, 16(1): 58.

Crampton, J.W. (2004) 'GIS and geographic governance: Reconstructing the choropleth map', *Cartographica*, 39(1): 41–53.

Crampton, J.W. (2007) 'The biopolitical justification for geosurveillance', *Geographical Review*, 97(3): 389–403.

Crampton, J.W. (2009) *Mappings: An Introduction to Critical Cartography and GIS*, Oxford: Blackwell.

Crampton, J.W. and Elden, S. (2007) *Space, Knowledge, and Power: Foucault and Geography*, Aldershot, UK: Ashgate.

Crampton, J.W. and Krygier, J. (2006) 'An introduction to critical cartography', *ACME: International E-Journal for Critical Geographies*, 4(1): 11–33.

DiBiase, D., DeMers, M., Johnson, A., Kemp, K., Luck, A.T., Plewe, B. and Wentz, E. (2006) *Geographic Information Science Body of Knowledge*, Washington, DC: Association of American Geographers.

Dominian, L. (1917) *The Frontiers of Language and Nationality in Europe*, New York: Henry Holt for the American Geographical Society.

Duster, T. (2005) 'Race and reification in science', *Science*, 307(5712): 1050–1.

Eicher, C.L. and Brewer, C.A. (2001) 'Dasymetric mapping and areal interpolation: Implementation and evaluation', *Cartography and Geographic Information Science*, 28(2): 125–38.

Elden, S. (2001) *Mapping the Present: Heidegger, Foucault, and the Project of a Spatial History*, New York: Continuum.

Elden, S. (2007) 'Governmentality, calculation, territory', *Environment and Planning D: Society and Space*, 25(3): 562–80.

Fall, J.J. (2005) *Drawing the Line: Nature, Hybridity and the Politics of Transboundary Spaces*, Aldershot, UK: Ashgate.

Foucault, M. (1977) *Discipline and Punish: The Birth of the Prison*, New York: Pantheon Books.

Foucault, M. (1978) *The History of Sexuality*, New York: Pantheon Books.

Foucault, M. (1984) 'Space, knowledge, and power', in P. Rabinow (ed.) *The Foucault Reader*, New York: Pantheon, 239–56.

Foucault, M. (1991) 'Governmentality', in C.G.G. Burchell and P. Miller (eds) *The Foucault Effect: Studies in Governmentality*, Chicago, IL: University of Chicago Press, 87–104.

Foucault, M. (2003) *Society Must Be Defended: Lectures at the Collège de France, 1975–76*, New York: Picador.

Foucault, M. (2007) *Security, Territory, and Population*, New York: Palgrave Macmillan.

Foucault, M. (2008) *The Birth of Biopolitics: Lectures at the Collége de France 1978–1979*, New York: Palgrave Macmillan.

Goodman, A.H., Heath, D. and Lindee, M.S. (2003) *Genetic Nature/Culture: Anthropology and Science Beyond the Two-Culture Divide*, Berkeley, CA: University of California Press.

Gould, S.J. (1977) 'Why we should not name human races—a biological view', in S.J. Gould (ed.) *Ever Since Darwin. Reflections in Natural History*, New York: W.W. Norton & Company, pp. 231–6.

Grant, M. (1916) 'The passing of the great race', *Geographical Review*, 2(5): 354–60.

Grant, M. (1917) 'Introduction', in L. Dominian (ed.) *The Frontiers of Language and Nationality in Europe*, New York: Henry Holt and Company, pp. xii–xviii.

Grant, M. (1932) *The Passing of the Great Race or the Racial Basis of European History*, 4th edn, New York: Scribner's Sons.

Graves, J.L. (2001) *The Emperor's New Clothes: Biological Theories of Race at the Millennium*, New Brunswick, NJ: Rutgers University Press.

Gregory, D. (1994) *Geographical Imaginations*, Cambridge, MA: Blackwell.

Grosz, E. (1995) 'Women, chora, dwelling', in S. Watson and K. Gibson (eds) *Postmodern Cities and Spaces*, Oxford: Blackwell, pp. 47–58.

Hacking, I. (1982) 'Biopower and the avalanche of printed numbers', *Humanities in Society*, 5: 279–95.

Häkli, J. (2007) 'Biometric identities', *Progress in Human Geography*, 31(2): 139–41.

Hannah, M. (2000) *Governmentality and the Mastery of Territory in Nineteenth-Century America*, Cambridge: Cambridge University Press.

Hard, G. (2001) 'Hagia Chora: Von einem neuerding wieder erhobenen geomantischen ton in der geographie', *Erdkunde*, 55(2): 172–98.

Harold-Smith, G. (1929) 'A three-dimensional thermo-isopleth', *Science*, 69: 404–5.

Heidegger, M. (1959) *An Introduction to Metaphysics*, trans. R. Manheim, New Haven, CT: Yale University Press.

Huxley, J. (1938) 'Clines: An auxiliary taxonomic principle', *Nature*, 3587: 219–20.

Jefferson, M. (1909) 'The anthropography of some great cities: A study in distribution of population', *Bulletin of the American Geographical Society*, 41(9): 537–66.

Kitchin, R. and Dodge, M. (2007) 'Rethinking maps', *Progress in Human Geography*, 31(3): 331–44.

Koch, T. (2005) *Cartographies of Disease: Maps, Mapping and Medicine*, Redlands, CA: ESRI Press.

Langford, M. and Unwin, D.J. (1994) 'Generating and mapping population density surfaces within a geographical information system', *The Cartographic Journal*, 31: 21–6.

Langford, M., Maguire, D.J. and Unwin, D.J. (1991) 'The areal interpolation problem: Estimating population using remote sensing in a GIS framework', in I. Masser and M. Blakemore (eds) *Handling Geographical Information: Methodology and Potential Applications*, New York: Wiley, pp. 55–77.

Legg, S. (2005) 'Foucault's population geographies: Classifications, biopolitics and governmental spaces', *Population, Space and Place*, 11(3): 137–56.

Lemke, T. (2001) '"The birth of bio-politics": Michel Foucault's lecture at the Collège de France on neo-liberal governmentality', *Economy and Society*, 30(2): 190–207.

Lewontin, R.C. (1972) 'The apportionment of human diversity', in M.K. Hecht and W.S. Steere (eds) *Evolutionary Biology*, New York: Plenum, pp. 381–98.

Lewontin, R.C. (1995) *Human Diversity*, New York: Scientific American Library.

Liddell, H.G. and Scott, R. (1990/1871) *A Lexicon Abridged from Liddell & Scott's Greek-English Lexicon*, Oxford: Clarendon Press.

Livingstone, F. (1962) 'On the non-existence of human races', *Current Anthropology*, 3(3): 279–81.

Lowenthal, D. (1969) 'Obituary: John Kirtland Wright 1891–1969', *Geographical Review*, 59(4): 598–604.

McCleary, G.F. (1969) *The Dasymetric Method in Thematic Cartography*, unpublished PhD thesis, University of Wisconsin–Madison.

MacEachren, A. (1979) 'The evolution of thematic cartography. A research methodology and historical review', *The Canadian Cartographer*, 16(1): 17–33.

McGushin, E.F. (2007) *Foucault's Askesis*, Evanston, IL: Northwestern University Press.

Marks, J. (1995) *Human Biodiversity*, Hawthorne, NY: Aldine de Gruyter.

Marks, J. (2006) *The Realities of Races*, mimeo. <http://raceandgenomics.ssrc.org/Marks/>.

Martin, D., Langford, M. and Tate, N.J. (2000) 'Refining population surface models: Experiments with Northern Ireland census data', *Transactions in GIS*, 4(4): 343–60.

Martin, S. (2005) *Cartography, Discourse and Disease: How Maps Shape Scientific Knowledge About Disease*, unpublished Masters thesis, Anthropology and Geography, Georgia State University, Atlanta.

Nash, C. (2003) 'Cultural geography: Anti-racist geographies', *Progress in Human Geography*, 27(5): 637–48.

National Academy of Sciences (2006) *Beyond Mapping: Meeting National Needs through Enhanced Geographic Information Science*, Washington, DC: The National Academies Press.

Newman, D. (2006a) 'Borders and bordering: Towards an interdisciplinary dialogue', *European Journal of Social Theory*, 9(2): 171–86.

Newman, D. (2006b) 'The lines that continue to separate us: Borders in our "borderless" world', *Progress in Human Geography*, 30(2): 143–61.

Olsson, G. (1998) 'Towards a critique of cartographical reason', *Ethics, Place and Environment*, 1(2): 153–5.

Olsson, G. (2007) *Abysmal. A Critique of Cartographic Reason*, Chicago, IL: University of Chicago Press.

Paasi, A. (2002) 'Territory', in J. Agnew, K. Mitchell and G. Toal (eds) *A Companion to Political Geography*, Oxford: Blackwell, pp. 109–22.

Painter, J. (2008) 'Cartographic anxiety and the search for regionality', *Environment and Planning A*, 40: 342–61.

Plato (1929) *Timaeus, Critias, Cleitophon, Menexenus, Epistles*, trans. R.G. Bury, Cambridge, MA: Harvard University Press.

Plato (1977) *Timaeus and Critias*, trans. D. Lee, London: Penguin Books.

Robinson, A.H. (1952) *The Look of Maps: An Examination of Cartographic Design*, Madison, WI: University of Wisconsin Press.

Robinson, A.H. (1971) 'The genealogy of the isopleth', *The Cartographic Journal*, 8: 49–53.

Robinson, A.H. (1982) *Early Thematic Mapping in the History of Cartography*, Chicago, IL: University of Chicago Press.

Robinson, A.H. and Wallis, H.M. (1967) 'Humboldt's map of isothermal lines: A milestone in thematic cartography', *The Cartographic Journal*, 4: 119–23.

Sallis, J. (1999) *Chorology: On Beginning in Plato's Timaeus*, Bloomington, IN: Indiana University Press.

Schuurman, N. (2009) 'Spatial ontologies', in R. Kitchin and N. Thrift (eds) *The International Encyclopedia of Human Geography*, Oxford: Elsevier.

Slocum, T.R., McMaster, R.B., Kessler, F.C. and Howard, H.H. (2008) *Thematic Cartography and Visualization*, Upper Saddle River, NJ: Prentice Hall.

Sluga, G. (2005) 'What is national self-determination? Nationality and psychology during the apogee of nationalism', *Nations and Nationalism*, 11(1): 1–20.

Smedley, A. (1999) *Race in North America. Origin and Evolution of a Worldview*, Boulder, CO: Westview Press.

Sui, D.Z. and Holt, J.B. (2008) 'Visualizing and analysing public-health data using value-by-area cartograms: Toward a new synthetic framework', *Cartographica*, 43(1): 3–20.

Tate, N.J., Fisher, P.F. and Martin, D. (2008) 'Geographic information systems and surfaces', in J.P. Wilson and A.S. Fotheringham (eds) *The Handbook of Geographic Information Science*, Malden, MA: Blackwell Publishing, pp. 239–58.

Turnbull, D. (2003) *Masons, Tricksters and Cartographers: Comparative Studies in the Sociology of Scientific and Indigenous Knowledge*, London: Routledge.

van Houtum, H., Kramsch, O. and Zierhofer, W. (2005) *B/Ordering Space*, Aldershot, UK: Ashgate.

Walker, F.A. (1874) *Statistical Atlas of the United States*, New York: J. Bien.

Walker, F.A. (1896) 'Restriction of immigration', *The Atlantic Monthly*, 77(464): 822–9.

Wallis, H.M. and Robinson, A.H. (1987) *Cartographical Innovations: An International Handbook of Mapping Terms to 1900*, London: Map Collector Publications for the International Cartographic Association.

Winlow, H. (2009) 'Mapping race and ethnicity', in R. Kitchin and N. Thrift (eds) *The International Encyclopedia of Human Geography*, Oxford: Elsevier.

Wright, J.K. (1938) 'Problems in population mapping', in J.K. Wright (ed.) *Notes on Statistical Mapping, with Special Reference to the Mapping of Population Phenomenon*, New York: American Geographical Society and Population Association of America, pp. 1–18.

Wright J.K. (1942) 'Map makers are human: Comments on the subjective in maps', *Geographical Review*, 32: 527–44.

Wright, J.K. (1944) 'A proposed atlas of diseases', *Geographical Review*, 34: 642–52.

3 Rethinking maps from a more-than-human perspective

Nature–society, mapping and conservation territories

Leila Harris and Helen Hazen[1]

Introduction

> Maps are not simply representations of particular contexts, places, and times. They are mobile subjects, infused with meaning through contested, complex, intertextual and interrelated sets of socio-spatial practices.
>
> (Del Casino and Hanna 2005: 36)

> Maps are of-the-moment, brought into being through practices (embodied, social, technical), *always* remade every time they are engaged with; mapping is a process of constant reterritorialization. As such, maps are transitory and fleeting, being contingent, relational and context-dependent. Maps are practices—they are always *mappings*, spatial practices enacted to solve relational problems . . .
>
> (Kitchin and Dodge 2007: 335)

This chapter interrogates the power dynamics of mapping practices and products from a more-than-human perspective. Specifically, we consider what is at stake in defining and mapping protected areas for conservation. We combine literatures related to critical cartographies, political ecology, and nature–society debates to shed light on the ways that conservation-mapping practices endorse certain notions of species, space, and territory—with profound implications for the ways that nature conservation is perceived and operationalized. We also turn to recent insights related to practices and engagements with "map spaces" (rather than reading maps as fixed representations of space) to consider conservation-mapping practices as dynamic, performative interactions among people, landscapes, ecosystems, and species. We argue that a focus on map spaces and practices provides useful ways forward for recent debates related to nature–society and more-than-human geographies. In turn, considering the more-than-human reveals further fruitful lines of enquiry related to critical cartographies and power effects of mapping.

Recently, there has been considerable attention to retheorizing maps; less as product, and more as practice (this volume; Del Casino and Hanna 2005). For instance, Kitchin and Dodge (2007) explain that cartography should be understood as processual, rather than representational. Questioning the ontological status of maps, they argue that maps are never fully formed, never complete, and are always brought into being through specific context-dependent practices and relations. As part of this refocus on processual cartographies they suggest that key insights are possible by analyzing the ways that lines and colours *become* maps, are given meaning, and are performed in relation to specific knowledges or techniques, or through relational engagements involving mapmakers or users. For instance, what are the technological or social codes that are cited and signified in ways that allow a reader to interpret and engage with the image as a map? What are the consequences and socio-spatial interactions that unfold in interpreting and engaging with the map? Similarly, Del Casino and Hanna (2005) write of the need to rethink mapping to consider multiple ways that maps are engaged and performed, particularly by map users. Rather than analyzing the map and its silences, or the power-knowledge dynamics at play that led to the creation and production of particular maps (e.g. imperialism, state mapping, or other power dynamics that might result in the production of maps; cf. Harley 1988, 1997), these authors argue that it is imperative to analyze the multiple, reiterative production and reproduction of maps as they are engaged in multiple times and spaces. Central to this rethinking, they highlight the need to overcome key dichotomies common to cartographic literatures, such as those of map user/maker, or representation/space.

The last decade of geographic scholarship has also witnessed a proliferation of discussions related to "animal geographies," and linked efforts to rethink human relationships to non-human or "more-than-human" natures (Braun 2005; Philo 1995; Philo and Wilbert 2000; Whatmore 2002, 2006; Wolch and Emel 1995, 1998). The term "more-than-human" is meant to move beyond the negative terminology of "non-human," and suggests the need for our interest, attention, and commitment to reach beyond an exclusive focus on the human world.[2] Lynn (2004) defines the term as also referring to human issues (e.g. environmental justice), while maintaining an appreciation for the theoretical and empirical linkages among animal, environmental, and human affairs. As examples of such concerns, theorists have analyzed spaces that are thought to be appropriate for diverse non-human animals, and the ways that particular spaces are defined in relation to non-human "others"; for example, policy and practice related to urban farming (Blecha 2007), acts of enclosure related to zoos or modern industrial agriculture (Watts 2000, 2004), or conceptualizations of "pests" in racialized urban spaces (Biehler

2007). These discussions touch on the ethical considerations of bounding and limiting animal spaces, and also make the case that social theory can benefit by paying greater attention to diverse non-human natures in discussions of power and socio-spatial interactions.

Our interest here is to combine insights related to critical cartographies of conservation with geographic discussions related to the more-than-human to analyze what these approaches together suggest for a critical reading of the common practice of "mapping for conservation." We therefore provide overviews of recent work in each of these literatures before considering how work at their intersection can serve to identify and challenge inequalities and inconsistencies in contemporary conservation practice. Given the important role of mapping in designating areas for conservation, and the resulting implications in terms of designating appropriate spaces for certain human activities and non-human species, what might a more explicit focus on mapping products *and* practices lend to these discussions? In particular, how might the processual turn in critical cartography enrich understandings of reiterative mappings, reterritorializations, and socio-spatial practices related to more-than-human geographies? In the conclusion, we also briefly address what more-than-human geographies might add to discussions of map spaces and performativities.

We use the terms "mapping for conservation" and "conservation cartographies" interchangeably to highlight a complex of interrelated spatial and territorial strategies common to contemporary conservation practice. These include: the designation of geographical areas as relevant for conservation, the delimitation of practices that are considered to be appropriate with respect to those areas, and cartographic representation and replication of those associations. Certainly, mapping for conservation also incorporates more than this, for instance habitat or species distribution mapping, or biodiversity assessments. However, this discussion relates most directly to the mapping, designation, and bounding of territorial conservation areas, such as parks or nature reserves. The explicit linking of conservation goals to specific territories is a practice that finds expression in an expanding map of protected areas (Harris and Hazen 2006; Naughton-Treves *et al.* 2005; Zimmerer *et al.* 2004), which today covers over 8 percent of the world's terrestrial surface (WDPA 2006), even by conservative estimates. As Woodley (1997: 11) suggests, the designation of protected areas has become "the most common human response to human induced ecosystem degradation."

Highlighting these connections, our key argument is that the relational or processual turn in cartography reveals important dimensions of power effects and dynamics of conservation mappings, offering key insights to nature–society debates in geography. We read conservation maps as practices that remake knowledges and truths related to the more-than-human world, reiteratively revisiting and cementing particular notions of what nature conservation is, and what it should be.

The processual turn in critical cartography

> There has been an increasing recognition that the image of the world can never fully represent the world, because the process of representing is itself part of the world being represented.
>
> (Perkins 2006: 556)

Kitchin and Dodge (2007) and Del Casino and Hanna (2005) are authors whose contributions are central to understanding what we refer to here as the "processual turn" in critical cartography. Among other points, these authors argue that we should not think of maps as "things" with ontological certainty. Rather, it is more useful to think of the relationships that unfold as maps are created, the meanings that are cited in selection of particular technologies or representational techniques by mapmakers, or as maps are engaged by users. Such an approach refuses to think of maps as static, or fixed, in terms of their meanings. Along these lines, Kitchin and Dodge (2007: 331) ask "what are the citational practices that are invoked by cartographers to produce something we recognize as a map?" Attention to this type of question, they argue, forces us to interrogate fundamental issues: What is a map? How do we recognize a map when we see it? What are the citational cues a mapmaker might invoke to sediment a particular notion of a map (drawing on a particular technological and aesthetic repertoire to produce something recognizable as such)? Not only are there key relationships that are engaged in the process of map production, but even once produced the map still does not exist in a stable, or ontologically-given, sense.

The approach these authors offer builds on earlier work that understands mapping as power-laden (per the work of Harley 1997; see also Crampton 2001; Pickles 1995). Elsewhere, we have applied insights from these discussions of the "power of maps" to evaluate conservation cartographies: What are the relevant power dynamics and asymmetries with respect to how conservation maps are drawn, by whom, and for what ends? (cf. Harris and Hazen 2006). Although this remains a solid foundation for our avenues of enquiry here, we consider that recent discussions related to critical cartographies force us to extend this type of analysis to acknowledge that maps are never stable, never fixed, and are constantly open to reinterpretation, and assignment of shifting meanings. Furthermore, if we break down clear separations between mapmakers and map users, we also need to be attentive to what this implies in terms of understanding power dimensions and effects of maps. In brief, we read these recent interventions as consistent with an interest in the power effects of mappings, but requiring that we deflect analytical attention away from the *intent* of the mapmaker (à la Harley) to instead consider the multiple, diffuse, and unpredictable ways that mapping practices and products are engaged and "remade." Among other elements of the ways that these contributions revise our interest in power of maps is attention to the ways that maps are read and invested with meaning in particular times and spaces. What is the power of the conservation map in terms of its representational and effective power, and how do diverse interpretations and power effects shift over time and across space? For example, how might conservation mappings and spaces in the global South be invested with different meanings from those in Northern contexts? Or how might conservation spaces and mappings hold particular meaning in relation to changing political and economic contexts? This reorientation builds on Pickles' interest in maps as "multivocal and contested," rejecting a singular notion of "truth" that can be uncovered by exposing ideological intent of the map's production (Kitchin and Dodge 2007: 333).

To clarify the implications of recent discussions for our reading of conservation mapping, it is worth providing more detail on this retheorization of maps associated with the processual turn. For instance, Del Casino and Hanna (2005) are particularly interested in the ways that spaces serve to condition map uses or meanings and, in turn, the ways that mappings may alter one's experience of space. In sum, they highlight spatio-temporalities of mapping as reiterative, co-constitutive and, indeed, power-laden. Del Casino and Hanna (2005: 44) argue that:

> Maps that people simultaneously make and use mediate their experience of space. People's bodily practices of walking, driving, touching, smelling, and gazing, as well as their understandings of landscapes and spaces can be guided and informed by maps and by the innumerable intertextual and experiential references always present in any map. At the same time, spaces mediate people's experiences of maps . . . our theorization therefore does not prioritise writing over reading or production over consumption in the constant recreation of the map space. Nor do we wish to argue that map spaces as representation are separable from map spaces as practiced, worked or performed.

This further clarifies how mappings can be power-laden beyond the intent or interest of the mapmaker. Instead, mapping products and practices are power-laden in serving to guide peoples' understandings and experiences of space and, in turn, spaces can serve to condition and render one's experience of a map. Therefore, we need to think of mapping practices as key aspects of socio-spatial dynamics and relations, whether socio-spatial exclusions, power relations, or differentiated experiences of spaces and places relevant for the more-than-human realm.

Highlighting power as key to their theorization, Del Casino and Hanna reference the work of Judith Butler to detail the "performative effects of maps." Reading maps and mappings as performative draws attention to the reiterative processes through which map meanings and effects are constantly remade. This analytic necessarily also draws attention to the ways that relations *appear* as stable or natural, even as they are constantly unfixed and remade. For instance, even if a map is "remade" with each reading, use, or engagement, there are still ways in which maps appear to cement or stabilize particular socio-spatial relations. This is consistent with discussions of maps in terms of their tendency to convey certainty and control, provide reassurance, or cement particular power dynamics (e.g. Perkins 2006). Particularly through the ubiquity of maps that allows them to appear so commonplace, everyday, and *a*political, there are key ways in which socio-spatial relations may appear as natural and stable, as effects of reiterative citation, even as the maps themselves, or their power effects, are not pre-certain, given, or fixed.

Taken together, we read recent contributions in critical cartography as refocusing and extending an interest in power dimensions of mapping. How might maps serve to naturalize certain relations—relations of power, particular political economic relations, or territorial partitioning—particularly given that maps are so authoritative, so everyday, and seemingly *a*political? Given that maps are reproduced in diverse spaces and times, how does attention to the spatiality and temporality of map production, uses, and engagements affect the condition and effects of particular maps and mapping practices? Further, how are maps central to understanding uses and experiences of space (again, particularly given their ubiquity, everydayness, and seemingly *a*political character)? Applied to the conservation context, in particular, how might mapping for conservation have power effects in terms of consolidating particular nature–society relations, and demarcating (in)appropriate spaces and relations for human:more-than-human interactions, even if these effects are not pre-given or fixed?

More-than-human perspectives and animal geographies

To adequately read conservation cartographies in this way also points to the need to extend common theorizations of power, enlivening an eco-social theorization of power that incorporates inequalities in inter-species and non-human senses.[3] Non-human animals are rarely considered within the realms of social theory (Wolch and Emel 1995, 1998), and yet Philo (1995: 655) argues that animals can be regarded as a "marginal 'social' group" that is "subjected to all manner of socio-spatial inclusions and exclusions." Consider, for instance, that many non-human species or dimensions of "life" itself are now frequently represented as "natural resources" (Whatmore 2006: 605), generating clear messages of consumption for human:non-human relationships. As we will detail, just as performativities of conservation mappings are necessarily linked to power relations that privilege certain social groups over others, similar hierarchies of values operate to privilege some species or non-living natures over others. For instance, charismatic species such as the panda are commonly used to mobilize funding and generate environmental concern (Lorimer 2007), and particularly biodiverse habitats such as rainforests are frequently seen as more significant conservation targets than are grasslands or other less spectacular landscapes (Hazen and Anthamatten 2004). How might we take these sorts of issue seriously in evaluating the performative and power effects of conservation mappings?

Attention to power effects of conservation mappings in eco-social senses also raises key questions related to modes of representation, in science and politics, of human and non-human (Braun and Castree 1998; Whatmore 2002). It becomes necessary to interrogate how the very notions of human:non-human are represented and articulated, and to what effects (Haraway 1991). These new "more-than-human" modes of enquiry "neither presume that socio-material change is an exclusively human achievement nor exclude the 'human' from the stuff of fabrication" (Whatmore 2006: 604). As populations or planners variously and reiteratively engage with conservation maps (in literal and metaphorical senses), how do such engagements cite, reconsider, challenge, or reify particular power relations between humans and non-human "others," solidify certain spaces as appropriate for particular species, or generate notions of "desirable" species that we seek to conserve? All of these types of questions illustrate the productive analytical space that is opened up by combining recent debates in critical cartography with animal geographies. It is to this analytical terrain, and to the particular concern of "conservation mappings" that is so central to these negotiations, that we now turn.

Critical cartographies and conservation geographies

Many conservation practices use mapping techniques and products. In particular, most contemporary conservation efforts rely on the designation of particular territories for conservation, resulting in complex mappings of spaces deemed appropriate, or inappropriate, for particular activities and human–nature relations. Most obviously, protected areas define areas where human activities are generally abstracted from, and secondary to, the needs of wildlife, although more recent notions such as buffer zones and mixed-use areas have complicated such simplistic separations of the human and non-human. As we have argued elsewhere (Harris and Hazen 2006; Hazen and Harris 2007), the designation of certain geographic areas for conservation, and reliance on mapping products and practice to do so, has several notable effects, including:

a   solidifying a notion that humans and non-human others are, and should be, separate (see also Fall 2002);
b   privileging those voices and perspectives that have access and expertise related to Western cartographic approaches and GIScience in conservation debates (see also Goldman 2003; Robbins 2003);
c   favoring those spaces, ecosystems, and natures that may be "more mappable" for protection over other areas (e.g. the fact that grasslands and marine areas are less definable in cartographic terms than forests and islands may help to explain why these features are relatively less well represented in protected areas);

d   cementing an overly-limited territorial approach to conservation, in ways
     that potentially sideline non-territorial approaches; and
e   consolidating an overly-fixed and static approach to conservation, rather
     than enabling approaches that might be more seasonal, fluid, or appropriate
     for shifting and evolving ecological conditions and needs (see also Clapp
     2004; Natter and Zierhofer 2002; Zimmerer 2000 for discussion of these
     issues).

As we have elaborated these issues elsewhere, here we simply offer a
few examples to re-examine these connections in light of the theoretical
interventions noted above.

Performative effects of conservation maps

Just as other theorists have considered the performative effects of maps in
producing nationalist sentiment, or consolidating state boundaries in ways that
allow them to appear natural, given, or ahistorical (e.g. Edney 1997; Radcliffe
and Westwood 1996; Winichakul 1994), it is clear that mapping particular
areas for conservation has similar productive effects. Here it is important to
clarify that we engage notions of performativity (following Butler 1990, 1993)
in a way that is distinct from certain other discussions of performative mapping
(e.g. discussions related to dancing or singing of spatial relationships, or ways
that mapping relates to other cultural performance; for instance, see Perkins
2006: 565 discussion of "acting out the map"). Instead, we draw on notions
of performativity from Butler as described in the introduction above. In this
theorization, maps are never fixed or stable, but are reiteratively engaged, with
particular effects. Butler's idea of performativity also draws centrally on the
idea of power effects as these reiterative engagements and performances are
necessarily understood as linked to power dynamics, whether particular
performances are "compelled," or help produce certain stabilities and fixities
of meaning (Butler 1990, 1993). Applied to conservation mapping, performativity helps underscore additional ways that maps may naturalize contingent
links between spaces and territories, and constellations of human–environment
relations generally. Though not engaging the same use of the term, this
discussion reveals features of what Harley (1988: 59) refers to in relation to
the silences of maps as "active performances" in terms of their "social and
political impact and their effect on consciousness." Drawing on this theorization,
we see key ways in which conservation mappings can be read as reiterative
performances that cite meaning and, in so doing, serve to consolidate particular
power effects vis-à-vis non-human and animal geographies.

Static, fixed, and inflexible associations

Perhaps foremost among the performative effects of mapping for conservation
practice is the tendency for map forms to provide static and fixed associations.
Just as reading a world political map may give a reader the sense that certain

political boundaries exist in ways that effectively erase dynamism inherent to the state system, maps may also provide snapshots of associations between ecosystem needs, species, and specific territories, despite tremendous flux and dynamism (both in terms of human-ecological systems and with respect to shifting scientific understandings). This is precisely what leads Clapp (unpublished mimeo) to note that conservation territories are a "blunt instrument unsuited for dealing with the natural world, characterized less by stability than flux in time and space." As Zimmerer (2000) argues, despite a shift towards non-equilibrium ecological understandings, many conservation principles continue to rely heavily on spatial parameters that "are premised almost entirely on equilibrium assumptions about the nature of environments" (p. 356). His critique of conservation areas "that apply, in rigid style, the ecological precepts of stable spatial boundaries, single scales, and the regular temporal quality of environments" (Zimmerer 2000: 357), is one that we share. Although creating fixed spaces of conservation may be sensible from the viewpoint of limiting uses of habitats that threaten certain species, all too often protected areas are taken as inviolable or fixed in space and time in ways that are fundamentally inconsistent with changing ecologies and species requirements. In terms of engendering fixity in terms of which areas or species should be protected, taking a more performative approach to mapping effects is illustrative. Consider, for instance, the example offered by Vandergeest (1996) in which mapped locations over time came to legally define "forests" for state agencies in Thailand, regardless of the actual vegetation in those spaces. This offers a powerful example of ways that maps can take on importance in and of themselves, potentially cementing fixed territorial associations, even if they have little bearing on ecological conditions or conservation requirements. Noting that spatio-temporal fixity of conservation areas is problematic is not to imply that conservation practitioners simplistically assume that specified boundaries remain effective over long time frames (see, for instance, Newmark 1995; Noss 2001; Shafer 1999) or that conservation territories are undesirable, but rather that their limitations should be understood. Further discussion of such challenges is clearly warranted.

A fundamental point we distill here is that mapping practices and products tend to cement inflexibility, making more flexible conservation approaches less visible and less likely. Related to this concern, conservation mappings may also reinforce the impression that conservation has *already* happened and is successful. The use of the past tense in the term "protected area" may be particularly significant in this regard, engendering a (false?) sense of security that conservation *has occurred* and is effective. The notion of "paper parks"—spaces protected in name but not in practice—is a clear example of this possibility. For instance, Cropper *et al.* (2001) found that in Thailand, although 10 percent of land is defined as "protected," the protection status of the land bears no statistically significant relationship to rates of forest clearing. Considerable recent evidence from biophysical and conservation literatures stress a wide range of extra-territorial threats to conservation spaces, including air and water pollution, climate change, and even water withdrawals, such as at the Everglades in the U.S. Simply put, designating an area as "protected" (in the past tense) and mapping it as such serves to cite links between species health and geographical territories in particular ways, often masking other extra-territorial and long-term threats to ecosystem health. Thus, although mapping conservation spaces may reiteratively cite the territorial basis for species and ecosystem health (causing other necessary links or threats to fade from view), the product and practice of conservation mapping may also overly-fix strategies for dealing with conservation (sidelining more flexible approaches), and even engender a sense of security that protected areas are successful (and that conservation has happened), potentially rendering other possible approaches less likely.

Mapping "Others"

One basic idea that frequently underwrites conservation mappings is that "animal geographies" can be mapped as separable "Others," cementing the idea that non-human species do not belong in urban, ex-urban, or even rural spaces unless they are set-aside as "wilderness" spaces. In this way, conservation maps cite and affirm particular roles and spaces for humans and non-humans, furthering the idea that there can, and should be, neat and separable boundaries between them. Spaces outside conservation areas are also important in this regard, as mapping areas for conservation may serve to justify intensified use and degradation of spaces outside of park boundaries (Zimmerer 2000). In this context, the reiterative power effects of conservation mappings can be read as consolidating certain landscapes and spaces of over-use, in contradistinction from spaces where nature conservation is prioritized, indicating as much about what occurs *outside* conservation spaces as *within* them. This notion of people and nature as separate is precisely what recent efforts attempting to re-integrate local populations in conservation spaces are attempting to overcome (e.g. Naughton-Treves *et al.* 2005).

As we have argued elsewhere, the idea that conservation goals can be achieved by setting aside spatially limited, biodiverse territories—the so-called "hotspots" approach—has become popular, partly because it may appear as more politically and economically feasible than more extensive conservation techniques (Harris and Hazen 2006; Hazen and Harris 2007). This is the case even as critiques suggest that the hotspots approach may not be sustainable in the long term, nor as effective as other strategies (see Kareiva and Marvier 2003 for critique of the hotspots concept). Hotspot conservation is perhaps an extreme example, but many practices and products of conservation mapping encourage conservation approaches that target limited geographical areas at the expense of other strategies. Returning to points made in the introductory section, there are often scalar or territorial references to administrative units, such as states, among citational practices that create maps. These underlying divisions, whether implicit or explicit, can condition subsequent mappings through influencing our expectations of what scale and extent are appropriate for particular conservation practices. For instance, the mapping of protected areas or nature preserves often happens at scales determined by administrative units and boundaries, even if these scales may be less appropriate to ecological or functional processes.

The limited territorial approach that mapping practices often implicitly endorse is particularly risky given the increasing acknowledgment that many species may not persist unless there are regional-, or even continental-, scale approaches to ensure their survival (e.g. Groves *et al.* 2000; Ricketts *et al.* 1999). For example, a report on grizzly bear conservation efforts in Yellowstone concluded that recovery of the park's bear population cannot occur in isolation from conditions throughout the continental U.S. (Willcox and Ellenberger 2000). Such examples offer explicit recognition that protected areas are typically too small and isolated to accommodate the needed movements and changes required by many species.

These insights reveal that conservation mappings may reiteratively produce an association between conservation and territory in a general sense, and conservation and particular scales of "nature" more specifically. However, conservation efforts may be appropriate at multiple scales, including not only the regional or even global scales, but perhaps also the molecular or genetic scale (a topic of increasing interest in nature–society geography and science studies—as evidence for this possibility Watts (2004) is one of several scholars who highlight the gene as central to the "modern map of nature").

Power effects of conservation boundaries: Shaping how humans and more-than-human communities experience space

We have discussed ways that conservation maps may consolidate overly fixed borders, in ways that might even undermine important ecological functions. Related to this, conservation maps play important roles in conditioning experiences of space, for humans and non-humans alike. For instance, the bison of Yellowstone experienced a particularly harsh winter in 2007–8; in response, many animals strayed outside the park boundary. Park authorities often slaughter bison that leave the park to prevent the spread of brucellosis to surrounding herds of cattle. This year the number of bison killed reached a new high—one-quarter of the park population— generating an outcry from critics of the current management system (Robbins 2008; see also Lavigne 2002 for general discussion of this issue). This example makes it clear that non-humans experience space in particular ways given the "mapped" park boundary, even as this boundary had no material expression "on the ground." The consequence of the bison transgressing the invisible mapped boundary of the park is especially severe, affecting the bison not only individually but also in terms of the genetic viability of the herd.

The symbolic significance of the bison, both for the U.S. West and for Native American tribes, also reveals some of the performative effects of conservation mappings. With each transgression of the bison over the border of the park, and with each individual brought to slaughter, each individual shifts from being "protected species" to "threat to political economic interests," and is resignified as "expendable" or "consumable." The performative effects of the map also invoke meanings and history related to the repeated loss of Native American livelihoods (re-enacted with each transgression of the map space and associated slaughter), and even notions of loss of pristine nature that have been central to the American imagination (Cronon 1995). Indeed, the effects of the map, given the contemporary political economic landscape (notably cattle grazing), reiteratively cite and perform each of these associations, histories, and nature–society relationships. Each transgression of the bison brings new meaning to the "map space" and power effects of the Yellowstone Park map, adding a compelling addition to our understanding of the stakes and consequences of "map spaces" for human and non-human relations and futures alike.

To provide another example of the ways that maps are performatively engaged in relation to space, consolidating particular understandings and uses of spaces, consider the production and replication of "protected areas" on print maps. A tourist looking at such a map may consider seeking out such spaces to have a "wilderness experience," or indeed may avoid such areas assuming that there will be nothing of interest for them. Thus, the map influences not only the map user's experience of space (as per Del Casino and Hanna's 2005 argument), but also alters patterns of mobility. Consider also how walking through an area demarcated as "wilderness" on a map might elicit a more (or, indeed, a less) satisfying "nature" experience as a result of expectations set by the map (again, with particular connotations in relation to particular cultural constructions of wilderness—see Cronon 1995). At a more abstract level, the production and dissemination of such maps could have the effect of affirming a basic idea that humans and non-human nature are, and should be, separate. Combined with other evidence of social exclusions related to protected-areas mappings (e.g. Goldman 2003; Peluso 1993), as well as biophysical evaluations of differential protection afforded to different ecosystems (Hazen and Anthamatten 2004), it is clear that protected areas are designated, mapped, and managed according to shifting notions of appropriate or desirable nature, as well as the priorities that societies, or certain subsets of society, deem to be important. Further, as Robbins (2001) has shown, remotely-sensed images, or any map of "nature," may be interpreted differently by users according to pre-existing notions or categories, further demonstrating the socio-spatial effects of conservation mappings.

Other performative effects: Social consciousness, erasure, and nature–society divides/inequalities

There are also clear ways that conservation maps can affect social and political consciousness by consolidating identities around environmentalism, nationalism, or wise-use movements. For instance, work in the U.S. context has highlighted the role of national parks in fuelling nationalist sentiment; with large parks fostering pride or facilitating belonging related to notions of "Americanness" (Hazen 2008). At another extreme, the mapping of a specific area as "protected" may fuel opposition to such practices, giving "wise-use" or similar movements ammunition for rejecting such examples of "government intrusion" that render land unavailable for local people (see McCarthy 2002 for discussion of this movement). The more classic example, perhaps, is that of mapping conservation spaces that promote ideals of "wilderness" in ways that erase complex histories of human settlement to comply with certain visions of pristine and uninhabited nature (as with the removal of Native American communities at Yellowstone National Park [cf. Spence 1999] or the clearing of pioneer settlements at Smoky Mountains National Park [see also Cronon 1995]). Ironically, even as parks of this type help to perpetuate the idea of people and nature as separate, once mapped in this way, many areas commonly associated with nature such as Niagara Falls and Yosemite actually undergo considerable human management and intervention to comply with idealized notions of wilderness or pristine nature.

As a final example along these lines, it is also worth highlighting the performative effects of maps in terms of the relative mappability of particular features or land areas. Given technological considerations, certain areas can be considered more "mappable" than others, and may as a result be privileged for conservation designation. Consider, for instance, that grasslands are not only considered less "majestic" than other landscapes (see Cronon 1995) but are also less definable in carto-geographic terms than, for example, lakes or islands, and may therefore be relatively neglected by conservation designations. The preference for the protection of forest over dryland and grassland ecosystems that can be seen at the global scale (Hazen and Anthamatten 2004) may also be, in part, a reflection of the fact that forests are often a "mapped" feature, whereas grasslands and drylands are invisible on all but the most specialized of maps.[4] This issue, again, reveals the reiterative and performative effects of mapping products and technologies in terms of producing particular landscapes and spaces deemed appropriate, or necessary for, protection. It is also of interest that many features of critical importance to the success of protected areas are "unmappable" in these terms (for example, social and political infrastructure), and thus may be neglected in a political climate that overemphasizes what can be achieved with innovative techno-logical tools or limited territorial set-asides.

Conclusions

Given the limitations of many common conservation-mapping practices—whether disregarding local knowledges or needs, fostering inflexible approaches that ignore changing conditions, or encouraging notions of conservation that justify separation between humans and nature—it is clear that there is compelling need for conservation cartographies to be debated, retooled, and remapped. Elsewhere, we outline practices and suggestions that offer some promise or partial solutions to the issues and challenges raised above, varying from ways to adapt the form and function of protected areas, to more fundamental rethinking of conservation spaces (Harris and Hazen 2006; Hazen and Harris 2007).

Here, we provide further examples of the ways that mapping practice and products may consolidate particular nature–society relations, potentially circumscribing a range of alternate futures that may be imaginable from a more-than-human perspective. Among the most important elements of the ways that mapping potentially limits and consolidates particular nature–society relations over others, we consider that mapping consolidates an *overly-fixed*, and relatively *inflexible*, range of possibilities with respect to animal spaces and futures. For instance, designating certain territories as "conservation areas," and enabling continued use and degradation outside of park boundaries, considerably limits future flexibility with respect to altering or extending protected-area boundaries or adjusting them in accordance with seasonal or long-term needs and requirements.

Understanding mapping as a common technology of conservation practice allows for more explicit interrogation of the spatial and territorial under-pinnings of conservation, as well as the limitations of common conservation mappings. The fundamental point that can be distilled from all of these divergent debates is that conservation maps, as any maps, are necessarily reflective of, and productive of, power. Just as we consider that attention to the processual turn in cartography lends considerable insights for an enriched understanding of the reiterative and performative effects of conservation mappings (with important implications for more-than-human possibilities and futures), we consider that critical cartographic discussions may also benefit from attention to the more-than-human world. It is not only the case, as other authors have noted, that engaging with maps may influence our sense of place, or help to guide our experience of spaces (Del Casino and Hanna 2000, 2005). This is certainly true. However, there are also considerable power effects that are revealed through explicit attention to the "experiences" of non-human others, as their use of space is conditioned, and circumscribed, by conservation mappings. Indeed, in many cases, the very survival of individuals or entire species is fundamentally linked to ways that conservation maps are reiteratively engaged by diverse users in diverse contexts, with particular effects.

Mapping products and possibilities are central to the contemporary conservation toolkit. They also frame and limit the specific geographies and management opportunities possible in terms of how human and more-than-human worlds are inhabited and lived. We have found inspiration in recent debates related to critical cartography to revisit some of our ideas about the role and effects of particular conservation techniques, approaches, and products. We similarly think that debates about power in cartography would do well to pay attention to power in its many diverse manifestations, human and more-than-human alike.

Notes

1   We are grateful to *ACME: International E-journal for Critical Geographies* and *Environmental Conservation* (Cambridge Journals) for allowing us to reproduce elements of earlier publications here. The full citations for those publications are listed in the References below as Harris and Hazen (2006) and Hazen and Harris (2007). We are also grateful for the work of the editors of this volume for their patience and assistance.
2   We use non-human and more-than-human interchangeably throughout this piece, even as we share a commitment to the spirit of this project.
3   More eco-social understandings of power have been elaborated elsewhere. For example, Sneddon (2007) identifies a "failure" with respect to conceptualizations of power, highlighting the fact that questions of ecology, or the non-human, are infrequently broached in discussions of power. Interestingly, as our discussion of power of conservation mappings extends from territorial and spatial assumptions in conservation practices, understandings of power in human geography are also often theorized in spatial terms, for instance the territorial expression of state power, or spatio-temporal ordering and discipline central to Foucault's (1982) understandings of governmentality and power/knowledge.
4   In a discussion related to some of the concerns of this paper, Vandergeest (1996) considers the mapping of forest areas as a critical step in the process whereby the Thai state asserted control over territory, people, and resources (eventually with the forestry department claiming control of nearly half of Thai national territory). Given such examples, it is clear that mapping practices are often central to the assertion of control and power (e.g. state power, see related discussions on surveillance in Foucault 1979 or on state legibility in Scott 1998). Further, it is suggestive that issues of control, surveillance, and resource access may be central to the determination of which features are preferentially mapped and protected.

References

Biehler, D. (2007) *In the Crevices of the City: Public Health, Urban Housing, and the Creatures we Call Pests 1900–2000*. Unpublished PhD thesis, Department of Geography, University of Wisconsin-Madison.

Blecha, J. (2007) *Urban Life with Livestock: Performing Alternative Imaginaries Through Small-scale Urban Livestock Agriculture in the United States*. Unpublished PhD thesis, Department of Geography, University of Minnesota.

Braun, B. (2005) 'Environmental issues: Writing a more-than-human urban geography', *Progress in Human Geography*, 29: 635–50.

Braun, B. and Castree, N. (1998) *Remaking Reality: Nature at the Millennium*, London: Routledge.

Butler, J. (1990) *Gender Trouble: Feminism and the Subversion of Identity*, London: Routledge.

Butler, J. (1993) *Bodies that Matter: On the Discursive Limits of "Sex"*, New York: Routledge.

Clapp, R.A. (2004) 'Wilderness ethics and political ecology: remapping the Great Bear Rainforest', *Political Geography*, 23: 839–62.

Crampton, J.W. (2001) 'Maps as social constructions: power, communication and visualization', *Progress in Human Geography*, 25: 253–60.

Cronon, W. (1995) 'The trouble with wilderness: or, getting back to the wrong nature', in W. Cronon (ed.) *Uncommon Ground: Toward Reinventing Nature*, New York: W.W. Norton & Company.

Cropper, M., Puri, J. and Griffiths, C. (2001) 'Predicting the location of deforestation: the role of roads and protected areas in Northern Thailand', *Land Economics*, 77: 172–86.

Del Casino, V.J. and Hanna, S.P. (2000) 'Representations and identities in tourism map spaces', *Progress in Human Geography*, 24: 23–46.

Del Casino, V.J. and Hanna, S.P. (2005) 'Beyond the "binaries": A methodological intervention for interrogating maps as representational practices', *ACME: An International E-Journal for Critical Geographies*, 4(1): 34–56.

Edney, M. (1997) *Mapping an Empire: The Geographical Construction of British India, 1765–1843*, Chicago, IL: University of Chicago Press.

Fall, J. (2002) 'Divide and rule: constructing human boundaries in "boundless nature"', *GeoJournal*, 58: 243–51.

Foucault, M. (1979) *Discipline and Punish: The Birth of the Prison*, New York: Vintage Books.

Foucault, M. (1982) 'Space, knowledge and power', in P. Rabinow (ed.) *The Foucault Reader*, New York: Pantheon.

Goldman, M. (2003) 'Partitioned nature, privileged knowledge: community-based conservation in Tanzania', *Development and Change*, 34: 833–62.

Groves, C., Valutis, L., Vosick, D., Neely, B., Wheaton, K., Touval, J. and Runnels, B. (2000) *Designing a Geography of Hope: A Practitioner's Handbook for Eco-regional Conservation Planning*, The Nature Conservancy, <http://conserveonline.org/docs/2000/11/GOH2-v1.pdf>.

Haraway, D. (1991) *Simians, Cyborgs and Women: The Reinvention of Nature*, New York, Routledge.

Harley, J.B. (1988) 'Silences and secrecy: the hidden agenda of cartography in early modern Europe', *Imago Mundi*, 40: 57–76.

Harley, J.B. (1997) 'Deconstructing the map', in J. Agnew, D. Livingstone and A. Rogers (eds) *Human Geography: An Essential Anthology*, Oxford: Basil Blackwell.

Harris, L. and Hazen, H.D. (2006) 'Power of maps: (Counter)mapping for conservation', *ACME: International E-Journal for Critical Geographies*, 4: 99–130.

Hazen, H.D. (2008) '"Of outstanding universal value": The role of the World Heritage Convention at national parks in the U.S.', *Geoforum*, 39: 252–64.

Hazen, H.D. and Anthamatten, P.J. (2004) 'Representation of ecological regions by protected areas at the global scale', *Physical Geography*, 25: 499–512.

Hazen, H.D. and Harris, L. (2007) 'Limits of territorially-focused conservation: a critical assessment based on cartographic and geographic approaches', *Environmental Conservation*, 34: 280–90.

Kareiva, P. and Marvier, M. (2003) 'Conserving biodiversity coldspots', *American Scientist*, 91: 344–51.

Kitchin, R. and Dodge, M. (2007) 'Rethinking maps', *Progress in Human Geography*, 31: 331–44.

Lavigne, J. (2002) 'Where the buffalo roam: boundaries and the politics of scale in the Yellowstone region', *GeoJournal*, 58: 285–92.

Lorimer, J. (2007) 'Nonhuman charisma', *Environment and Planning D: Society and Space*, 25: 911–32.

Lynn, W.S. (2004) 'Animals: a more-than-human world', in S. Harrison, S. Pile and N.J. Thrift (eds) *Patterned Ground: Entanglements of Nature and Culture*, London: Reaktion Press.

McCarthy, J. (2002) 'First world political ecology: lessons from the wise use movement', *Environment and Planning A*, 34: 1281–302.

Natter, W. and Zierhofer, W. (2002) 'Political ecology, territoriality and scale', *GeoJournal*, 58: 225–31.

Naughton-Treves, L., Holland, M.B. and Brandon, K. (2005) 'The role of protected areas in conserving biodiversity and sustaining local livelihoods', *Annual Review of Environmental Resources*, 30: 219–52.

Newmark, W.D. (1995) 'Extinction of mammal populations in western North American national parks', *Conservation Biology*, 9: 512–26.

Noss, R.F. (2001) 'Beyond Kyoto: forest management in a time of rapid climate change', *Conservation Biology*, 15: 578–90.

Peluso, N. (1993) 'Coercing conservation? The politics of state resource control', *Global Environmental Change*, 3: 199–217.

Perkins, C. (2006) 'Mapping', in I. Douglas, R. Huggett and C. Perkins (eds) *Companion Encyclopedia of Geography*, London: Routledge.

Philo, C. (1995) 'Animals, geography and the city: notes on inclusions and exclusions', *Environment and Planning D: Society and Space*, 13: 655–81.

Philo, C. and Wilbert, C. (2000) *Animal Spaces, Beastly Places: New Geographies of Human–Animal Relations*, London: Routledge.

Pickles, J. (1995) *Ground Truth*, New York: Guilford Press.

Radcliffe, S. and Westwood, S. (1996) *Remaking the Nation: Place, Identity and Politics in Latin America*, London: Routledge.

Ricketts, T.H., Dinerstein, E., Olson, D.M., Loucks, C.J., Eichbaum, W., DellaSala, D., Kavanagh, K., Hedao, P., Hurley, P.T., Carney, K.M., Abell, R. and Walters, S. (1999) *Terrestrial Ecoregions of North America: A Conservation Assessment*, Washington, DC: Island Press.

Robbins, J. (2008) 'Anger over culling of Yellowstone's bison', *The New York Times*, March 23, <www.nytimes.com/2008/03/23/us/23bison.html>.

Robbins, P. (2001) 'Fixed categories in a portable landscape: the causes and consequences of land-cover categorization', *Environment and Planning A*, 33: 161–79.

Robbins, P. (2003) 'Beyond ground truth: GIS and environmental knowledge of herders, professional foresters and other traditional communities', *Human Ecology*, 31: 233–53.

Scott, J.C. (1998) *Seeing Like a State*, New Haven, CT: Yale University Press.

Shafer, C.L. (1999) 'National park and reserve planning to protect biological diversity: some basic elements', *Landscape and Urban Planning*, 44: 123–53.

Sneddon, C. (2007) 'Nature's materiality and the circuitous paths of accumulation: dispossession of riverine fisheries in Cambodia', *Antipode*, 39: 167–93.

Spence, M.D. (1999) *Dispossessing the Wilderness: Indian Removal and the Making of the National Parks*, New York: Oxford University Press.

Vandergeest, P. (1996) 'Mapping nature: territorialization of forest rights in Thailand', *Society and Natural Resources*, 9: 159–75.

Watts, M. (2000) 'Afterword: enclosure', in C. Philo and C. Wilbert (eds) *Animal Spaces, Beastly Places: New Geographies of Human–animal Relations*, London: Routledge.

Watts, M. (2004) 'Enclosure: a modern spatiality of nature', in P. Cloke, P. Crang and M. Goodwin (eds) *Envisioning Human Geographies*, New York: Arnold.

WDPA (2006) *Growth in Naturally Designated Protected Areas (1872–2006)*, World Data of Protected Areas, UNEP <www.unep-wcmc.org/wdpa/PA_growth_chart_2007.gif>.

Whatmore, S. (2002) *Hybrid Geographies: Natures, Cultures, Spaces*, Thousand Oaks, CA: Sage.

Whatmore, S. (2006) 'Materialist returns: practising cultural geography in and for a more-than-human world', *Cultural Geographies*, 13: 600–9.

Willcox, L. and Ellenberger, D. (2000) *The Bear Essentials for Recovery: An Alternative Strategy for Long-term Restoration of Yellowstone's Great Bear*, Sierra Club Grizzly Bear Ecosystems Project <http://sierraclub.org/grizzly/reports.asp>.

Winichakul, T. (1994) *Siam Mapped: A History of the Geo-body of a Nation*, Honolulu, HI: University of Hawaii Press.

Wolch, J. and Emel, J. (1995) 'Bringing the animals back in', *Environment and Planning D: Society and Space*, 13: 632–6.

Wolch, J. and Emel, J. (1998) *Animal Geographies: Place, Politics and Identity in the Nature-culture Borderlands*, London: Verso.

Woodley, S. (1997) 'Science and protected area management: an ecosystem-based perspective', in J.G. Nelson and R. Serafin (eds) *National Parks and Protected Areas: Keystones to Conservation and Sustainable Development*, Berlin: Springer-Verlag.

Zimmerer, K.S. (2000) 'The reworking of conservation geographies: nonequilibrium landscapes and nature-society hybrids', *Annals of the Association of American Geographers*, 90: 356–69.

Zimmerer, K.S., Galt, R.E. and Buck, M.V. (2004) 'Globalization and multi-spatial trends in the coverage of protected-area conservation (1980–2000)', *Ambio*, 33: 520–9.

4   Web mapping 2.0

Georg Gartner

Introduction

The term Web 2.0 was initially popularized by O'Reilly Associates in 2004 to reflect changes in the ways in which the World Wide Web was being deployed, and has subsequently come to stand for what is a potentially revolutionary change in the nature of the Internet. Web 2.0 extends the traditional Web by employing an architecture of participation that goes way beyond following hyperlinks. In this next generation of networked services a website is used as a platform for others to extend or edit content or services, instead of simply disseminating information created by a webmaster. Examples of Web 2.0 applications include social networking sites, video sharing and podcast sites, wikis, blogs and folksonomies. Such websites are designed to work in a social, collective and participatory manner, as is the open source software that underpins their development. Such software is increasingly being used in the development of cartography and mapping services, and a number of Web 2.0 mapping applications are active across the Internet.

Web mapping in Web 2.0 applications differs significantly from the first generation of Web mapping. Web 2.0 applications include collaborative, volunteer-led base map compilation such as OpenStreetMap and hybrid Web publishing that relies upon feeds, blogs, wikis and especially mashups. These new ways of using the Web alter the way in which information is accessed: users often become map producers and assemble data from many discrete and dispersed sites.

In this chapter the main issues of Web mapping 2.0 are discussed, as well as the consequences for cartographers and users. Questions over the quality, integrity, design and aesthetics, privacy and potential influences of governments or commercial companies are key for the success of the mapping in Web 2.0. It is argued that Web mapping 2.0 enables the integration of social and technical aspects into models of cartographic communication and that the process of technological change is itself leading to an important rethinking of mapping.

Web 2.0 and Web mapping 2.0

Web 2.0

The concept of Web 2.0 began with a conference brainstorming session between O'Reilly and MediaLive International in 2004. Although the term suggests a new version of the World Wide Web, it does not refer to an update to any technical specifications, but instead was coined to reflect changes in the ways software developers and end-users deploy the Web. The core competencies of the concept of Web 2.0, as coined by O'Reilly (2005) were as follows:

1   the Web as a platform with cost-effective scalability instead of packed software;
2   harnessing collective intelligence: using the 'wisdom of the crowds'[1] (e.g. as a filter for incorrect or inaccurate information);
3   control over unique, hard-to-create data sources that get richer as more people use them (serving as a provider of data and tools instead of maps; an architecture of participation);
4   trusting users as co-developers: including rich user experiences that would otherwise not be accessible;
5   leveraging the long tail through the consumer self: reaching out to the entire Web, to the edges and not just the centre, via user networks;
6   lightweight programming models: simplicity and organic Web-based, open source software, with no or little intellectual property protection, designed for 'hackability' and remixability;
7   software above the level of a single device: seamless connection of new devices (e.g. mobile devices) to the platform.

Web 2.0 websites allow users to do more than just retrieve information, encouraging them to add value to the application as they use it. According to Best (2006), the characteristics of Web 2.0 are: a rich user experience, user participation, dynamic content, metadata, Web standards and scalability. Three further important characteristics are openness, freedom (Greenemeier and Gaudin 2008) and collective intelligence (O'Reilly 2005) by way of user participation. This user participation depends upon the gradual development of what has been termed the semantic Web.

Berners-Lee *et al.* (2001: 3–4) define the semantic Web as 'an extension of the current Web in which information is given well-defined meaning, better enabling computers and people to work in cooperation'. This extension consists of metadata describing the semantics of the Web pages in a machine-processable way. Metadata are always represented by the Resource Description Framework (RDF), which is based on subject-predicate-object triples. Before Web pages can be described with semantic metadata, an ontology has to be defined for the domain or discourse. An ontology formally describes concepts found in the domain, the relationships between these concepts, and the properties used to describe them. W3C defines two ontology languages for the semantic Web: RDF Schema and the Web Ontology Language (OWL) and research by Jorge Cardoso (2007: 24) suggest, 'the language with the strongest impact in the semantic Web is without a doubt OWL'. Berners-Lee *et al.* (2001) proposed a Semantic Web Stack (Figure 4.1) that includes seven layers: Unicode+URI; an XML+NS+XML Schema; RDF+rdfschema; an Ontology vocabulary; Logic; Proof; and Trust. Subsequent research on the semantic Web has been related to these seven layers.

The sometimes complex and continually evolving technological infrastructure of Web 2.0 is built upon these conceptual schemas and includes server-software, content-syndication, messaging-protocols, standards-oriented browsers with plugins and extensions, and various client-applications. Web 2.0 websites typically include some of the following features/techniques: CSS, Folksonomies (collaborative tagging, social classification, social indexing and social tagging), Rich Internet application techniques (Ajax), syndication, aggregation and notification of data in RSS or Atom feeds, mashups (merging content from different sources, client- and server-side), weblogs and wikis.

In summary, there is no single, simple definition for Web 2.0 and there is no single, new technology that is driving its development. Rather a plethora of new ideas and applications are generating a shift in the meaning and use of the Web. The significance of Web 2.0 can be seen in the textual associations of the term mapped out in Figure 4.2, and in particular in the changed designs, economy, convergence, participation, usability, standardization and

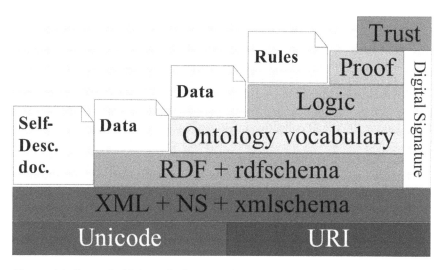

Figure 4.1 Semantic Web stack. Source: Tim Berners-Lee, Semantic Web – XML2000, www.w3.org/2000/Talks/1206-xml2k-tbl/slide10–0.html.

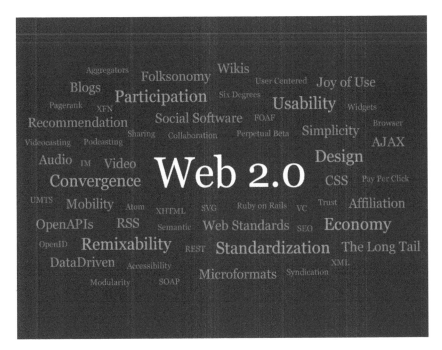

Figure 4.2 Textual associations to Web 2.0. Source: www.aperto.de, created by Markus Angermeier, http://nerdwideweb.com.

remixability that are facilitated by these changes. Rather than understanding this development as a change of paradigms (Kuhn 1962), it can be seen as an evolving technology. The main changes associated with these shifts are in the variability of content, enhanced interactivity and collaborative nature of Web 2.0.

Web mapping 2.0

The term Web mapping 2.0 is used here to refer to Web 2.0 applications that have a spatial frame of reference. Possible applications include: search engines considering spatial distance to find results; geotagging (virtually referring to objects in real space or on maps); geoblogging (enhancing blogs or photos with spatial references); and Web mashups (combining map data in a collaborative way). Mashups in particular characterise the technological changes associated with Web 2.0 and are the most popular new form of mapping associated with Web 2.0.

A mashup is a Web application that combines data from more than one source into a single integrated tool, but is much more than a simple embedding of data from another site to form a compound document. Content used in mashups is typically sourced from a third party via a public interface or API

Web service, with the data processed in some way so as to increase its value to users. Mashup APIs are used to connect different information (feeds) to the geospatial Web. These data are encoded in data formats such as GeoRSS or KML. Mashups currently come in three general types: consumer mashups (combining data elements from multiple sources, hidden behind a simple unified graphical interface); data mashups (mixed data of similar types from different sources); and business mashups (combinations of all the above, focusing on data aggregation and presentation, and additionally adding collaborative functionality, making the end result suitable for use as a business application).

The architecture of mashup Web applications is always composed of three parts: the content provider (acting as the source of the data, and making them available using an API and different Web-protocols such as RSS, REST, and Web Service); the mashup site itself (the Web application that provides the new service using different data sources); and the client Web browser (the user interface of the mashup). For example, Google acts as the content provider for many Google Maps mashups, providing geographically referenced image and map data, a standardised user interface and an API (cf. Gibson and Earle 2006). Mashup sites use the Google Map API to link data to the Google backdrops and design their own user interfaces to allow the mashed data to be accessed by users. For example, http://the-2008-olympics.blogspot.com allows you to get inside the Olympic Games with Google Maps, but without the advertisements that bedevil commercial news feeds about sporting stories. It maps medal tallies, events and stadia as these change.

Web mashups can be used to integrate different data sources, but in order to solve the semantic problems of data integration (such as scaling and naming conflicts) an ontological approach sourced in work on the semantic Web has to underpin the implementation. Ontology-based information integration is critical for Web mapping: the semantic heterogeneity of multiple information inevitably leads to data integration conflicts that can only be properly resolved by designs informed by semantic approaches (Wache *et al.* 2001). The widespread and continuing popularity of Web mapping 2.0 mashups reflects serious research in this area and Table 4.1, sourced from a sample of recent mashup tags, indeed shows that mapping is the most dominant mashup application.

One of the reasons for the dominant position of mapping as the most popular mashup applications can be found in the popularity of Google Maps as top application programming interface (API). Table 4.2 shows that nearly half of all the mashups sourced in the programmableweb.com website deploy the Google Maps API.

The preponderance of Google-based mashups indexed in tags is mirrored in the popularity of the Google Maps API in the same sample: Table 4.3 shows its disproportionate popularity among APIs in terms of the numbers of different mashups using each parent data source. According to the above figure, the three most popular APIs for mapping are Google Maps, Microsoft

Table 4.1 Top mashup tags, August 2008 (%)

Mapping	38
Photo	10
Shopping	9
Search	8
Video	7
Travel	6
News	4
Social	4
Messaging	4
Sports	4

Source: http://programmableweb.com.

Table 4.2 Top APIs for mashups, August 2008 (%)

API	Popularity
Google Maps	47
Flickr	11
YouTube	9
Amazon	7
MS Virtual Earth	4
eBay	4
411Sync	3
Yahoo Maps	3
Del.icio.us	3
Yahoo	3

Source: http://programmableweb.com.

Table 4.3 The 12 most popular APIs, August 2008

API	Service	Number of mashups
Google Maps	Mapping	1,497
MS Virtual Earth	Mapping	153
Yahoo Maps	Mapping	117
Yahoo Geocoding	Geocoding	77
GeoNames	Geocoding	43
Geocoder	Geocoding	24
Multimap	Geocoding	12
Yahoo Map Image	Map image creation	10
BigTribe	Location-based advertising	10
geocoder.ca	Geocoding	7
Poly 9 Freeearth	Three-dimensional mapping	7
MS MapPoint	Mapping	7

Source: http://programmableweb.com.

Virtual Earth and Yahoo Maps. These APIs are each free for use and provide similar functionality. Data formats for geographic annotation are available for all of those APIs, the most popular are GeoRSS and KML.

To summarise: Web 2.0 and Web mapping 2.0 applications depend upon many different technologies but mashups using APIs are the most popular and are easy to design and use. These mashups enable the integration and visualization of different geographic information on base map (such as Google Maps, Microsoft Virtual Earth or Yahoo Map). Consequently, technology is no longer a restricting factor when developing Web mapping 2.0 applications.

The significance of Web mapping 2.0

It is well recognized that mapping has perhaps changed more profoundly in recent years than at any time in the past. Among the main components contributing to these changes are the application of the computer to cartography (including the production of continuously updated and animated representations) and remote sensing – imagery produced through permanent surveillance of the earth by satellites (Thrower 1999). In the past, map production was limited to professional cartographers, and dissemination was radial: users were amateurs who consumed professional products. The revolutions in digital cartography described by Monmonier (1985) did little to alter expert control over mapping and GIS, if anything exacerbated the power of the producer.

Nowadays, however, websites such as Google Earth offer everyone the chance to produce their own individual maps, in many cases without the need of any professional qualification. Never before has this democratisation been as widely spread. So many users today are brought together via the Internet, that producers and consumers are no longer distinguishable. These developments evoke ambivalent attitudes: some have argued they spell the end of traditional cartography (e.g. Wood 2003). However, many traditional cartographic principles remain valid in the era of Web cartography, and mapping served on many websites still conforms to the more directional dictates of first-generation Web mapping (sometimes referred to as 'Web 1.0'), in which a large number of users are able to view contents provided by a comparatively small number of sites. Many of these maps on the Web are of the static, view-only type, originally sourced from scanned paper maps and designed for presentation. There is however an increasing need for dynamic and particularly interactive maps (Kraak and Brown 2001; Peterson 2003) and Web 2.0 offers a suitable platform for these new approaches for acquiring, assembling and publishing geographic information.

Others embrace a new era of map production with undreamed-of possibilities offered by the bi-directional collaboration that characterises the new mapping worlds of Web 2.0. New terminologies are emerging to reflect this changing conceptual field: in contrast to the old cartography, 'neo-geography' is emerging, embracing the integrative power of ubiquitous computing in which geography brings together the formerly separate (Turner

2006). Others cast these developments as volunteered geographic information (VGI), stressing the potential of locally sourced or volunteered information (cf. Goodchild 2007). Goodchild (2007) specifies three 'levels of sophistication': volunteered gazetteers produced entirely by individual citizens who add rich descriptions of places and hyperlinks to specific locations (e.g. Wikimapia, www.wikimapia.com); projects in which volunteers contribute substantial technical content, as in the OpenStreetMap project (www.openstreetmap.org); and those that allow contributors to make their own comparatively complex information available to others within easy-to-use Web 2.0 environments, such as Google Earth (http://earth. google.com/). This collaborative and social aspect of new mapping has also led others to liken the changes to the crowd-sourced and shared creation of online resources exemplified above all by the development of online encyclopaedia Wikipedia. In this view mapping too is becoming 'wikified' (Sui 2007).

The potential consequences of applying Web 2.0 principles to cartographic practice pertain to several overlapping topics. Information may be accessed in different ways in Web 2.0 mapping. Web cartography now offers wide possibilities to anybody who uses the Internet. Retrieval of geographic information ranges from classical forms of static maps, through to more interactive, dynamic and animated maps. In this respect, the Web offers many interesting new features (multimedia, hyperlinks, etc.), but it also has its limitations (Kraak and Brown 2001). For instance, Internet users may well have shorter attention spans; information may take too long to download; designs may be unattractive, or only poorly functional; and the quality of information is sometimes doubtful. Crowd-sourced mapping in Web 2.0 applications is only ever as good as the wisdom of the particular crowd that collects the data. There may be six billion potential local contributors in the global crowd (Goodchild 2007), each of whom has a unique geographical knowledge of familiar areas and many of whom may be eager to share their experiences, but we still have to face the fact that access to the Internet and required skills to use Web mapping are geographically and socially very uneven. An enormous amount of people will never have the chance to publish their geographic knowledge on the Web and many will not want to share their knowledge with strangers. Hence, without denying the obvious chances provided by enhancing Web maps with the geographical expertise of local residents, it is clear that some areas will remain comparatively uncharted territory. Moreover, mechanisms have to be developed to ensure quality and detect and remove errors, in order to achieve the same level of trust that traditionally produced maps enjoy.

Moreover, Web information is restricted to those who have access to appropriate skills to use the relevant software. Many of the Web 2.0 applications rely upon new kinds of specialist knowledge, in which a new cadre of 'experts' controls the technology and which are used by a large array of largely passive viewers. It has also been argued that Web 2.0 developments are largely driven by commercial concerns (cf. Scholz 2008). Google own the API and display their mapping backdrops for commercial reasons. They could remove both. They encourage mashups so as to be able to use volunteered data for their own commercial agendas. So it could be argued that Web 2.0 applications merely replicate digital divides, instead of subverting or offering radical change.

There are also debates around the role of the user as a map producer. Users have turned out to be very interested in building their own private maps and visualising their individual spatial data. An enormous number of people are willing to spend a lot of time contributing, without any hope of financial reward (cf. Haklay *et al.* 2008). Despite the fact that the quality of mapping that they produce varies and the fact that many do not meet cartographic standards, the trend towards user-generated maps and geographic information is probably inexorable and further development of easy-to-use and efficient tools and reliable data sources promises to improve individual contributions.

Web 2.0 certainly offers the potential for new things to be mapped for the first time. Spatial knowledge is hidden in many small segments of information fragments such as addresses on Web pages, annotated photos with GPS coordinates, geographic mapping applications, or geotags in user-generated content. Assembling data from these dispersed sites offers a very great potential in Web 2.0 mapping applications and can lead to a plethora of new maps. As the Web becomes more location-aware so geographical information science will have to resolve the best ways of bringing this disparate information together. Boll *et al.* (2008) argue that this process will involve increasingly interdisciplinary work.

As users naturally bring in different levels of skills in producing maps and geographic information, so must issues of reliability and data quality become more important. Displaying the reliability of both data sources and producer would be highly beneficial for consumer confidence. Existing metadata may be inappropriate in this context. New models for maintaining trust and 'rating' shared information already exist in the world of social networking and Web 2.0. Sellers on eBay rate other sellers; Facebook allows users to report inappropriate behaviour; Wikipedia edits inappropriate content and so on. Mechanisms for maintaining geographic trust need to be established in the world of Web mapping 2.0, and including rules for resolving 'crowd conflict'.[2]

Designing maps for the Web is a very challenging task, as the opportunities and limitations offered by on-screen maps in general, and specific characteristics of the Web in particular both have to be considered: usability engineering of sites is likely to become increasingly important (van Elzakker *et al.* 2008). Professional cartographers are just beginning to identify design principles for the Web in order to achieve effective and aesthetic digital maps. They sometimes find it hard to adjust their map-design habits. However, private users/producers are not necessarily tied to existing orthodoxies and can adopt ingenious alternatives, instead of following sometimes hackneyed, stereotyped or inappropriate fashion. Although this may not always result in admissible or even readable maps, some ideas could be useful. In this respect, crowd-sourcing the improvement of editable online maps can facilitate the identification of user-oriented design principles.

Concerns around privacy and users' rights to their own data are a serious challenge in the world of Web 2.0, especially as companies have begun to realize one of their chief sources of competitive advantage is controlling private data. Open data projects must therefore allow users to take control of their data, and there are potential conflicts between the shared nature of many sites, in which data are created according to the terms of Creative Commons licensing, and the much more private and controlled needs of the commercial sector. These developments are strangely paradoxical. Web users seem to be prepared to share and publish large amounts of personal and private information over the Web, as evidenced in the proliferation of blogs, personal websites, travel diaries of globetrotters (www. wherethehellismatt.com), and popular social networks such as MySpace or Facebook, but are strongly concerned about keeping information about themselves private or secret. This paradoxical willingness to share and reveal, while protesting the need to maintain privacy, seems to be an important characteristic of the new media. Torkington (cited in O'Reilly 2005) notes that:

> It's interesting to see how each Web 2.0 facet involves disagreeing with the consensus: everyone was emphasizing keeping data private, Flickr/Napster/*et al.* make it public. It's not just disagreeing to be disagreeable (pet food! online!), it's disagreeing where you can build something out of the differences. Flickr builds communities, Napster built breadth of collection.

There is a significant concern over Google Street View displaying geo-referenced photographs, but a wide willingness to submit georeferenced information to newly available map mashups.

In addition to these practical implications however, there are important conceptual considerations underlying the shift to Web 2.0 and it is to these that we now turn.

Cartographic communication and the semiotics of Web mapping 2.0

In the period after World War II cartography increasingly came to be understood as a communication process, where the transmission of spatial information could be explained as a process of coding and submitting that code to a user, who then decoded this information by reading the map. The goal of this process was seen as enabling the user to acquire spatial knowledge for orientation, navigation, planning, spatial analysis or other practical purposes.

During the twentieth century academic cartographers developed a number of conceptual schemas for understanding this process. The most well known of these is the map communication model, which uses a signal communication metaphor in which cartographic information is encoded via various filters (scale, projection, selection, etc.) into the map itself (the signal), and then decoded via further user-based filters (viewing time, map expertise, prior knowledge of the subject and so one; cf. Kolacny 1970). Subsequent research on contemporary cartography has often focused on developing this model in cognitive, semantic or theoretical approaches (cf. Ogrissek 1987). Significant impacts on the understanding of cartography as a form of communication have been provided by Bertin (1974), Gould and White (1974), MacEachren (1995) and Robinson and Petchenik (1975).

However, Freitag (2008) argued that all of these models lacked an appreciation of the social context of communication. In order to cover all aspects of cartographic communication he proposed that a model needed to integrate dialogue-oriented processes with monologic and collaborative communication. He also argued that the function of maps for defined user groups has to be made explicit, so that the communication model could be relevant to the concrete actions of users (cf. Dransch 2004). Such a model would define cartography in a more holistic way and encompass social aspects often neglected in research previously dominated by technicians.

It can be argued, that within the cartographic communication models, communication of meaning depends upon a number of different dimensions of what might be termed cartographic semiosis (cf. Jobst 2008; Wolodtschenko 2003). Derived from general communication models (Lidov 1999) these dimensions can be differentiated into syntactical, semantic and pragmatic aspects. Syntactics is the branch of semiotics that deals with the formal properties of signs and symbols; semantics deals with the meaning of the symbols; and pragmatics deals with all the psychological, biological and sociological phenomena that surround the functioning of cartographic signs.

Each influences communication, so understanding, addressing or even controlling those dimensions is key for the success of every communication process. The syntactical dimension can be explored to explain how graphical codes should be defined and whether they can be clearly perceived. The semantic dimension can be used to explain the methods being used to design maps as a whole, so they are efficient, useful and practical. The pragmatic dimension relating signs to users and their actions, has however, until recently, been rather neglected in cartographic research. As this deals primarily with psychological and sociological aspects it is difficult to incorporate into formal communication models. However, actions of human beings in the real world have to be an essential part of all theoretical models of cartographic communication processes, and the technological changes associated with Web 2.0 make it much more pressing to focus upon pragmatics.

To date, most academic cartographic research has focused on controlling the syntactical dimension of communication, and trying to provide better maps by exploring useful semantic aspects of semiosis. The new technical possibilities offered by Internet cartography and the first generation of Web mapping had limited impacts on semiotic academic cartographic research. However, the collaborative and participative nature of Web mapping 2.0 will lead to a change in research priorities. Pragmatics is likely to receive much more attention. It is the user's behaviour and interests that determines the communication process in Web 2.0: semantics and especially symbol and sign syntax are usually beyond the control of collaborative users. The increasing significance of the user in contemporary mapping systems is reflected in the newly established International Cartographic Association commission on use and user issues, and increasing research is being devoted to the contextual studies (see the theme issue edited by van Elzakker *et al*. 2008). The question remains open, however; whether modern cartography will be able to react to this challenge by offering (automated) methods and techniques, which help to define syntactical and semantic issues as well (Figure 4.3).

So cartography in the age of participative mapping will be challenged to define and offer rules, methods and techniques that can be applied to the collaborative data input. Topics such as cartographic generalisation, cartographic symbolisation and visualisation techniques will all be profoundly altered in the world of Web 2.0. Cartographers are needed to engineer frameworks to design syntactical and semantic dimensions of systems and for the first time in cartographic history the pragmatic dimensions of cartographic communication can be properly addressed in scientific research.

Conclusion

In this chapter the technologies underpinning Web 2.0 have been outlined, and the nature of Web mapping 2.0 has been described. The implications of

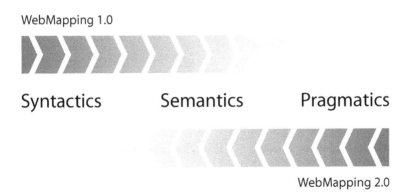

Figure 4.3 Change of research emphasis in the semiotic dimensions.

these shifts for cartographic practice and the impacts on rethinking cartographic research have also been evaluated. Key issues that must be resolved in order to release the full potential of Web mapping 2.0 include: how users can contribute in participatory systems; the availability of standardised programming tools and languages in open-source software communities; the design of new Web mapping applications; and (especially) research agendas investigating syntactically correct and semantically useful maps in pragmatic contexts that are collaborative and participative.

Web 2.0 offers great possibilities in the field of cartography, promising to open up new vistas of acquiring, assembling and publishing geographic information and forcing academics to change the ways in which they think of the map and mapping. Professional producers of maps and geographic information can benefit from these developments. The scientific project WikiVienna, for example, uses photographs of the City of Vienna contributed by anyone taking pictures with a camera or camera phone to build a collaborative three-dimensional model of the city.[3] Professional Web cartographers, and users producing their own maps, should therefore not be regarded as competitors. Instead, Web 2.0 offers us all a chance to enhance presentations of geographic information alongside traditional forms of maps. A user comment on an online discussion concerning Web mapping puts it this way: 'Well, I know we are not quite there yet, but on the other hand, isn't it a weird situation where strangers do maps about my area? I mean, who knows a region better than a local?'[4]

Notes

1   This phenomenon is also dubbed crowd-sourcing – using a crowd of consumers as creators (cf. www.businessweek.com/innovate/content/jul2006/id20060713_755844.htm).
2   See http://wiki.openstreetmap.org/index.php/Disputes/ for recent procedures established by the OpenStreetMap project for handling disputes and 'edit wars'.
3   See www.vrvis.at/research/projects/wikivienna/index.html.
4   See http://highearthorbit.com/cartographic-perspectives-on-the-doom-of-web-mapping/#comments.

References

Berners-Lee, T., Hendler, J. and Lassila, O. (2001) 'The semantic Web', *Scientific American Magazine*, 17 May.
Bertin, J. (1974) *Graphische Semiologie*, New York: De Gruyter.
Best, D. (2006) *Web 2.0 Next Big Thing or Next Big Internet Bubble? Lecture Web Information Systems*, Eindhoven: Technische Universiteit Eindhoven.
Boll, S., Jones, C., Kansa, E., Kishor, P., Naaman, M., Purves, R., Scharl, A. and Wilde, E. (2008) 'Location and the Web (LocWeb 2008)', *Proceeding of the WWW 2008*, 21–25 April 2008, Beijing, China, pp. 1261–2, <http://www2008.org/papers/pdf/p1261-locweb.pdf>.

Cardoso, J. (2007) 'The semantic Web vision: where are we?', *IEEE Intelligent Systems*, September/October: 22–6.

Dransch, D. (2004) 'Kartographie und Handlungstheorie' in W. Koch (ed.) *Theorie 2003*, Dresden: Kartographische Bausteine.

Freitag, U. (2008) 'Von der Physiographik zur kartographischen Kommunikation', *Kartographische Nachrichten*, 58(2): 59–67.

Goodchild, M.F. (2007) 'Citizens as voluntary sensors: Spatial data infrastructure in the world of Web 2.0', *International Journal of Spatial Data Infrastructures Research*, 2: 24–32.

Gibson, R. and Earle, S. (2006) *Google Map Hacks*, Sebastopol, CA: O'Reilly.

Gould, P. and White, R. (1974) *Mental Maps*, Harmondsworth, UK: Pelican Books.

Greenemeier, L. and Gaudin, S. (2008) 'Amid the rush to Web 2.0, some words of warning', *InformationWeek*, 26 May, <www.informationweek.com/shared/printableArticle.jhtml?articleID=199702353>.

Haklay, M., Singleton, A. and Parker, C. (2008) 'Web mapping 2.0: The neogeography of the geospatial internet', *Geography Compass*, 2(6): 2011–39.

Jobst, M. (2008) *Ein Semiotisches Modell für die Kartographische Kommunikation mit 3D*. Unpublished PhD Thesis, Technical University of Vienna.

Kolacny, A. (1970) 'Kartographische Informationen – ein Grundbegriff und Grundterminus der modernen kartographie', *Internationales Jahrbuch für Kartographie*, 10: 186–93.

Kraak, M-J. and Brown, A. (2001) *Web Cartography: Developments and Prospects*, London: Taylor & Francis.

Kuhn, T. (1962) *The Structure of Scientific Revolutions*, Chicago, IL: University of Chicago Press.

Lidov, D. (1999) *Elements of Semiotics*, New York: St. Martin's Press.

MacEachren, A.M. (1995) *How Maps Work*, New York: Guilford Press.

Monmonier, M. (1985) *Technological Transitions in Cartography*, Madison, WI: University of Wisconsin Press.

Ogrissek, R. (1987) *Theoretische Kartographie*, Gotha: Haack.

O'Reilly, T. (2005) *What Is Web 2.0 – Design Patterns and Business Models for the Next Generation of Software*, <www.oreillynet.com/pub/a/oreilly/tim/news/2005/09/30/what-is-web-20.html>.

Peterson, M.P. (2003) *Maps and the Internet*, Amsterdam: Elsevier.

Robinson A.H. and Petchenik, B.B. (1975) 'The map as a communication system', *The Cartographic Journal*, 12(1): 7–15.

Scholz, T. (2008) 'Market ideology and the myths of Web 2.0', *First Monday* 13(3), <www.uic.edu/htbin/cgiwrap/bin/ojs/index.php/fm/article/view/2138/1945>.

Sui, D. (2007) 'Volunteered geographic information: A tetradic analysis using McLuhan's law of the media', *Workshop on Volunteered Geographic Information*, 13–14 December, <www.ncgia.ucsb.edu/projects/vgi/products.html>.

Thrower, N.J.W. (1999) *Maps & Civilization: Cartography in Culture and Society*, Chicago, IL: University of Chicago Press.

Turner, A. (2006) *Introduction to Neogeography*, Sebastopol, CA: O'Reilly Short Cuts.

van Elzakker, C., Nivala, A-M., Pucher, A. and Forrest, D. (2008) 'Caring for the users', *The Cartographic Journal*, 45(2): 84–6.

Wache, H., Vögele, T., Visser, U., Stuckenschmidt, H., Schuster, G., Neumann, S. and Hübner, S. (2001) *Ontology-Based Integration of Information A Survey of*

Existing Approaches, IJCAI–01 Workshop Ontologies and Information Sharing, 2001.

Wolodtschenko, A. (2003) 'Cartography and cartosemiotics: Interaction and competition', *Proceedings of the 21st International Cartographic Conference (ICC)*, Durban, South Africa, 1976–98.

Wood, D. (2003) 'Cartography is dead (thank god!)', *Cartographic Perspectives*, 45: 4–7.

5 Modeling the earth

A short history

Michael F. Goodchild[1]

Introduction

Models are partial mirrors, reflecting how humans perceive the world around them. I trace the history of modeling the geographic world from early description and representation in the form of drawings and stick maps to today's geographic information systems. Representations have grown more complex and detailed with time, but no less partial in the sense that their content tells only part of the story, and never tells any part exactly. I argue that representation is cyclical: an approach is adopted in response to some kind of technological advance; it is found to be partially successful, and is adapted and stretched to include phenomena for which it was not designed; and finally as stresses build a new approach is adopted that exploits advances in technology. I examine the current situation in this context, and argue that the cycle will repeat itself before long.

Mapping is almost as ancient as humanity itself, and so also are the terms that people use to describe various aspects of the earth's surface—nouns and adjectives for features, prepositions to express relationships, and verbs to describe change. A variety of tools have been developed to aid in description, and to create representations that can be stored, shared, and changed as knowledge advances. Among the earliest of these were the sticks used to etch map-like diagrams on mud floors and the sticks and string used by Pacific islanders to aid in navigation. By the time of the invention of the printing press the idea of a map had become codified, along with the concepts of projection and symbolization that were necessary to reduce knowledge of the earth's surface to a collection of marks on flat paper. With a pen and paper, a cartographer could sketch the outlines of continents, add rivers and roads, and annotate with feature names. The continuous variation of topography presented more of a problem, but eventually contour lines became the conventional way of expressing change in elevation. In essence traditional pen-and-paper cartography evolved as a way of coding the features of the world in symbolic form.

Like many other areas of human activity, the various stages of map-making from data acquisition to compilation and editing and eventual distribution have become computerized, and now take advantage of many of the benefits of the digital world: transmission at electronic speed, automated numerical calculation, and easy copying and editing. People have learned how to express geographic knowledge in the binary alphabet required by digital computers, and have adopted a number of standard procedures that allow individuals separated by large distances and by differences of culture and language to communicate that knowledge effectively.

Essential to these advances are the various rules used to express geographic knowledge in binary code—the equivalent for the geographic world of music's MP3, and the digital world's replacement for the conventions of pen and paper. Several distinct phases can be identified in the development of these digital representations, or *data models* to use the conventional technical term, each achieving a significant advance on the previous one by being able to capture and express a wider range of geographic phenomena with greater fidelity. However, no representation can possibly be perfect, since the real world is infinitely complex. Over the past two decades much attention has been focused on the concept of uncertainty, or the degree to which a representation leaves its users uncertain about the true nature of the real world (Zhang and Goodchild 2002). There are many sources of uncertainty, including measurement error, vagueness in the definitions of terms, and the need to force information into the template provided by a data model. Most advances in representation and in the technologies of data acquisition provide a closer approximation to the real world, but the ideal of perfect representation is necessarily unachievable.

This chapter provides a brief history of the digital representation of geographic information, and the advances that have been made. Although each advance is an improvement on its predecessors, removing constraints, expanding the set of geographic information types that can be represented, and reducing inherent uncertainty, the goal of a comprehensive representation of all geographic information remains distant. The next three sections describe the major phases, each following from the adoption and adaptation of improved data modeling concepts from the computing mainstream. This is followed by an assessment of the current state of the art, represented by the object-oriented paradigm, and of recent advances in data-modeling research. A cyclical model is presented, in which new adoptions are followed by increasing stress, as geographic information types for which the adopted approach is essentially inappropriate are nevertheless forced into its template, and finally by replacement with a new, more powerful approach. The chapter ends with speculation about the prospects for a new cycle.

Three phases

Flat files

One claimant to the title of first geographic information system (GIS) is the Canada Geographic Information System (CGIS), developed by the

Government of Canada with the assistance of IBM in the 1960s. It was motivated by the Canada Land Inventory, a massive effort to map the capabilities and current use of land within a large swath of Canadian territory. The primary objective of the inventory was to provide statistics on the current and potential use of land, and the amounts of land available for new uses. This required a large mapping effort, followed by a detailed analysis based largely on the measurement of area.

Within this domain it was possible to argue that all maps were essentially of the same type: they showed land divided into irregularly shaped areas, each with homogeneous characteristics. For example, the maps of soil capability for agriculture divided geographic space into areas of uniform capability, measured on an ordinal scale and qualified by numerous codes. Even today, the majority of mapping of soils, surface geology, land use, land cover, vegetation type, and habitat are of this type (Figure 5.1), which Mark and Csillag (1989) term the *area-class map*, though other terms such as *polygon coverage* are also in use. More formally, such maps can be regarded as depicting functions that map location **x** to one of a set of k classes, $c = c(\mathbf{x})$, in other words a nominal field.

At the time, the only practical medium for large-volume digital storage was the magnetic tape, a linear structure that required substantial tape movement to access records in other than their stored order. The designers of CGIS faced a significant problem: how to represent the contents of an area-class map as a linear sequence of binary digits. This could clearly be done by scanning, if a suitable device could be developed that would convert

Figure 5.1 An area-class map; part of the CalVeg map for the Santa Barbara area, showing areas of approximately homogeneous vegetation type. Source: author.

the maps to raster representation, and if the cells of the raster could be represented row by row using some appropriate scheme to convert the brightness of the map in each cell to a set of binary digits. But the requirements of CGIS and the interests of efficient performance seemed to indicate a different approach that would focus on the boundary lines on the map. In the *vector* approach a line is represented as a series of points connected by straight segments, creating what is known as a *polyline*. A complete patch is represented by a connected sequence of such polylines, forming a polygon.

Thus one might consider representing an area-class map as a set of polygons, each composed of straight segments connecting points, the contents of each being described by a set of coded attributes. In the terminology of mainstream computing such a solution would create a *flat file*; a collection of records of the same type. Such records would be of variable length, because polygons would vary in shape complexity and thus in the numbers of points required to code them, but their representations could be laid end to end on magnetic tape.

In practice this solution turns out to be quite inconvenient, because it results in every internal boundary of the map being represented twice, resulting in unnecessary duplication. Moreover, if the data are edited or transformed in any way, it is difficult to ensure that the two versions of each internal boundary remain identical. Instead, the designers of CGIS adopted a solution that is far from intuitive but results in much-improved performance. In this solution the basic record is not the polygon, but the section of common boundary between two polygons. Each such record contains descriptions of the polygons on each side, along with the points required to describe the common boundary's geometry. Various terms have been assigned to these polylines, including *arcs*, *chains*, *edges*, and *links*, the first being responsible for half of the name of the most popular commercial GIS software, ArcInfo. In CGIS these records formed a flat file, and were recorded in sequence on magnetic tape. Many of the analyses needed by CGIS, such as measurement of polygon area, turn out to be more efficiently performed on arcs rather than polygons, even though the arc-based solution contains no direct representation of the latter.

This flat-file solution worked well for CGIS, which despite initial teething problems was by 1975 in full production (Foresman 1998). Maps of current land use, soil capability, or capability for recreation could all be expressed in the same basic form as area-class maps. Systems that came later adopted the same basic approach, which could also be applied to maps of administrative areas, census reporting zones, or forest stands; each of these aspects of the geographic world could be represented as a collection of common boundaries between areas of homogeneous characteristics. It was possible to represent holes and islands, in other words polygons unconnected to the rest of the boundary network, if imaginary "causeways" could be inserted to connect them to the network, and ignored during analysis or visualization (such causeways are easily recognized because the same polygon appears on both

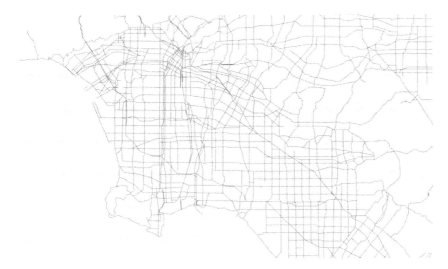

Figure 5.2 Major roads of the Los Angeles basin. It was possible to handle dead ends by extending the basic model devised for area-class maps. Source: author.

sides). Road networks could be represented if one relaxed the requirement that every arc end in two junctions (Figure 5.2), though overpasses and underpasses presented an additional problem. Data sets with complex attributes, such as those that result from the reporting of census summaries, were also a problem since attributes had to be repeated for every arc surrounding a polygon. When attributes changed or were edited it was difficult to ensure that every copy remained identical.

The arc-based solution adopted by CGIS was elegant, but it nevertheless did some damage to the truth as represented by the source map, which in turn simplified the real world. Although the source map could show curved boundaries, these had to be rectified into polylines in the digital system. Moreover, the map's use of homogeneous areas to describe what were essentially continuous variations of phenomena over the earth's surface added to the uncertainty inherent in the representation. Much effort has been expended in devising ways of representing this uncertainty. The *confusion matrix*, for example, records comparisons between the characteristics found at points on the earth's surface with the characteristics recorded for those points in the data, and summarizes the table in simple statistics such as Kappa (Stehman 1996).

The relational model

In the 1970s a new data model emerged in the computing mainstream that was to dominate thinking for the next decade. Tape storage was slowly giving way to disk, which allowed randomly selected records to be retrieved without the delays involved in winding tape. The relational model exploits this random-access ability, and is based on the following assumptions:

1   All information can be expressed in the form of tables or *relations*, each row defining one record, object, or case, and each column defining one of the characteristics of each record, object, or case. Clearly one can make such a table of the polygons of an area-class map, using one row for each polygon and one column for each of the known attributes of the polygon. However, polygons have a variable number of coordinates, so the table concept is not suitable for describing each polygon's geometric shape.
2   A given database can include many tables, recording the properties of different types of objects. Tables are linked through common keys or pointers; for example, a record in a table of patients might be linked to (point to) a record in a table of doctors, or a record in a table of passengers might be linked to a record in a table of flights.
3   The database can be *normalized* through a series of formal steps designed to ensure that no information is duplicated unnecessarily. For example, in an airline reservation system the information that details flights is stored not with the passenger records but in a separate table of flight records. In this way changes to flight details can be made once, rather than by editing the records of every passenger taking the flight.

The design adopted by ESRI for its first version of ARC/INFO represented a very significant step beyond the flat files of CGIS, by exploiting the relational model. The INFO database management system, one of the early commercial products to implement the relational model, was used to create a much more powerful solution to the problem of representing area-class maps. Two linked tables were created in what became known as the coverage model (Figure 5.3):

1   A table of polygons, recording the attributes of each one as a series of columns.
2   A table of arcs, each one linked to two polygon records by common keys, one pointing to the polygon on the left of the arc and one to the polygon on the right (left and right are defined by the order in which the arc's sequence of points is stored). Arc records also included pointers to the nodes or junctions at each end of the arc, though these nodes did not have their own table in early versions.

Because the structure stores explicit representations of the relationships between arcs, polygons, and nodes it is often described as *topologically rich*, or simply *topological*. But the coordinates needed to define each arc's geometry could not be stored in the structure, and instead were stored in a

Polygon table

ID	Name
A	Jasper
B	Newton
C	Pocahontas
D	Greenbrier

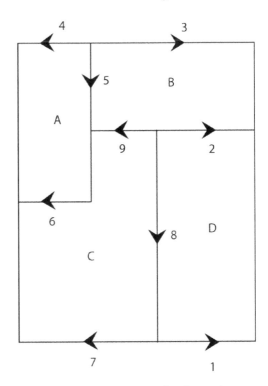

Arc table

ID	Left	Right
1	D	
2	B	D
3		B
4	A	
5	B	A
6	C	A
7		C
8	D	C
9	C	B

Figure 5.3 An example of a topological structure. The map contains four polygons and nine arcs. The tables display the contents that would be stored in a relational-model representation, with pointers from arcs to the polygons of which they are part. Source: author.

uniquely structured file outside the relational model. This solution, which was adopted by ESRI and is reflected in the choice of ARC/INFO as the name of the software, has been termed the *hybrid* model (DeMers 1997) for this reason. Almost three decades later, far more powerful and flexible relational database management systems are able to accommodate a much wider range of data types in the cells of their tables, including complex representations of polylines and polygons, and the need for the hybrid solution has largely disappeared.

Although this adaptation of the relational model, or *geo*-relational model, could clearly accommodate area-class maps, it almost immediately came under pressure from applications involving other geographic data types that did not fit the area-class model. As noted earlier, it is possible to fit road networks into the model provided that special allowances are made for cul-de-sacs, which end at otherwise unconnected nodes (nodes of valency 1). Moreover, the model cannot distinguish between overpasses, underpasses, and intersections at grade, since nodes must exist at every crossing. This latter problem was eventually addressed by creating a *turntable*, and explicitly enumerating all of the turns that could be made at any crossing. Transportation applications also require the ability to position landmarks, accidents, and other point-like events at points on arcs that may not be nodes. This was addressed by the concept of *dynamic segmentation*, which allowed features to be located using a system of *linear addressing* based on identifying the relevant arc and the distance along it from its starting node.

Several further extensions were made to the basic model. Data about points could be handled by allowing nodes that were not connected to the boundary network (nodes of valency 0), or by allowing polygons of zero area. Changes to a boundary network through time could be handled by creating a single database of all boundaries that ever existed, and adding a *region* data set that defined how polygons should be constructed from these primitive pieces at any specific date. In this way it was possible to accommodate the changing boundaries of the U.S. census, easements on property, overlapping wildfires, and many other complex phenomena.

In short, in the two decades following the adoption of the geo-relational model a series of extensions were made that accommodated new applications, but at the same time created an increasingly unmanageable superstructure. The software needed to handle the extensions became increasingly complex, terminology became increasingly confusing, and GIS software began to resemble a house of cards, with hundreds of basic concepts and thousands of commands.

By contrast, the approaches being used in computer-assisted design (CAD) software were far simpler and easier to understand. GIS software that adopted this simpler approach began to compete in the marketplace, putting pressure on the mainstream GIS software industry to find similarly simple approaches. When ArcView appeared in the mid 1980s it was marketed as an easy-to-learn entry to GIS, with the hope that its users would eventually migrate to the more powerful ARC/INFO. Its data-model template recognized features on the earth's surface as points, lines, or areas, respectively represented digitally as points, polylines, and polygons. No topological relationships between these features were accommodated, however, since the software was intended only for visualization of data, and no support was provided initially for the operations that benefit from topological structure: digitizing, editing, and analysis.

In time, however, pressure built to add such capabilities, and at the same time computer power and storage capacity grew, making it no longer imperative to avoid the double internal boundaries of area-class maps, and making it possible to compute topological relationships as and when required. ESRI also departed from past practice by publishing the format of its ArcView files (the shapefile format).

In summary, the early 1990s were characterized by a proliferation of data models. The original topologically rich coverage model had become increasingly compromised by numerous extensions, while the shapefile model had provided the user community with a comparatively simple alternative. Some vendors had attempted to replace this complexity with a simpler, more uniform solution, but none had achieved significant market share. Conditions were ripe for a new approach that would sweep away much or all of the complexity, and introduce a more coherent data model grounded in a better understanding of the nature of geographic data.

Object-oriented modeling

Mainstream thinking about data modeling had advanced a long way in the two decades following the popularization of the relational model. Object-oriented data modeling offers a more comprehensive approach that accommodates several concepts missing from the relational model, based on the following assumptions (Zeiler 1999; this discussion has been adapted to the particular needs of GIS):

1   All things, cases, events, instances, or objects of interest can be placed into classes.
2   Every member of a class can be distinguished on the same set of characteristics or attributes.
3   Classes can be specializations of more general classes and *inherit* their properties.
4   The members of a class can be aggregations of members of other classes, or composed of members of other classes.
5   The members of a class can be related to members of other classes through associations.
6   Methods can be associated or *encapsulated* with classes.

For example, all of the 50 states of the U.S. are members of the class *state*. States have numerous distinguishing attributes, including area, name, population, and date of admission into the Union. States are polygons, and have all of the characteristics of polygons (area, perimeter length), and also have more specific properties that polygons in general do not have (population, name). States are composed of counties, and have many associations; for example, an association exists between U.S. cities and their containing states (every U.S. city lies in exactly one state, although a state may contain any number of cities including 0).

The object-oriented approach proved to be far more powerful and flexible than the geo-relational model. It introduced the hierarchical concepts of inheritance, aggregation, and composition that had been entirely absent in the earlier solution, and allowed GIS database designers to capture the essentially hierarchical nature of many geographic phenomena, from the administrative hierarchies of township–county–state–nation to the scale hierarchies of river and road networks. It provided a general framework for many of the problems that had previously been addressed through special extensions: regions, dynamic segmentation, and temporal change. It allowed topological relationships between objects to be handled much more flexibly, and it allowed editing rules and other constraints to be represented as methods encapsulated with classes.

ESRI introduced object-oriented modeling in 1999 in ArcInfo Version 8, and later merged ArcView with ArcInfo into a single product line. Just as with the introduction of the relational model from the computing mainstream in the early 1980s, object-oriented modeling provided a uniform solution to an accumulation of problems, and a fresh, more powerful start on the representation of geographic phenomena. The basic classes of GIS—polygons, polylines, points—were specialized in a series of major GIS application domains, allowing users in those domains to access a standard template that had been designed to accommodate all of the classes commonly encountered in that domain (Arctur and Zeiler 2004). For example, the UNETRANS data model developed for transportation applications includes specializations of polygons to Traffic Analysis Zones and specializations of polylines to rail lines, bicycle paths, and canals. Object-oriented models have been constructed for applications that have never been associated with maps, helping to move GIS further away from its dependence on the map metaphor (Goodchild 1988), and towards a comprehensive approach to the representation of all types of geographic information.

Remaining issues

Despite the success of object-oriented data modeling, there are reasons to believe that the story is not yet finished, and that a new round of innovation may be needed. Two arguments lead in this direction, both grounded in the fundamental realities of the geographic world. Both arise as objections to the first assumption of object orientation given above, with its implication that all geographic phenomena can be conceptualized as things, events, cases, or objects—in other words as discrete. This assumption is as fundamental as the nature of computing itself.

Nevertheless there are numerous phenomena on the earth's surface that are fundamentally continuous, and for which discretization, or the breaking of phenomena into discrete pieces, is to some degree inappropriate or problematic. Roads, for example, are continuous, but are typically broken into pieces at their intersections and represented as collections of discrete polylines. Rivers are similarly segmented at junctions, or at real or imagined breaks, into reaches. More generally, it has long been recognized that humans approach the geographic world in two distinct ways (Longley *et al.* 2005). In the first, the *discrete object view*, the metaphor of an empty tabletop is used to conceptualize the world as empty except where it is occupied by discrete, countable objects that may or may not overlap and may or may not cover the table. Biological organisms, vehicles, and buildings fit this model well. On the other hand, other phenomena are better conceptualized as *continuous fields*, in which every location **x** is mapped to a single property *z* that may be nominal, ordinal, interval, or ratio; and scalar, vector, or tensor. Properties such as elevation, ownership, wind speed and direction, soil class, and current land use fit this model well.

To handle representations of continuous fields in GIS it is first necessary to discretize their continuous variation, using one of a number of approaches. Six of these are commonly implemented in GIS, while many others are in common use in specific scientific domains. But the resulting discrete objects are indistinguishable from collections that represent phenomena conceptualized as discrete objects. The result is inconsistency, complexity in the user interface, and an environment in which it is easy to make errors. Longley *et al.* (2005) use the example of eight points, which might represent eight cities or eight weather stations, but are otherwise indistinguishable as GIS data sets. In the second case, which implies a continuous-field conceptualization, it is reasonable to apply methods of spatial interpolation to estimate atmospheric properties in the spaces between the stations. But in the first case, which implies a discrete-object conceptualization, it would clearly be absurd to interpolate a property such as population count.

In the world of object-oriented data modeling these issues should in principle be handled by methods encapsulated with each class. Thus if a set of polylines represents digitized elevation contours, a method should prevent any edit that will result in contours crossing. Similarly a method should be associated with a set of polygons representing states to prevent any edit resulting in an overlap or gap between adjacent states, since the property *state* is conceptualized as a nominal field with exactly one value at every point within the national boundary. Two types of polygons should be distinguished, one related to discrete objects and the other to continuous fields; and in the second case all specializations should inherit the appropriate methods. Similar strategies should be adopted with respect to polylines and other classes that can be used to discretize fields. To date, however, no such implementations have been described.

Although the field/object distinction is powerful and covers a wide range of phenomena, there is now a recognition that other phenomena may not fall neatly into either category. Cova and Goodchild (2002) describe *object fields*, in which every location in space–time maps to an entire object, and show how viewsheds, watersheds, and trade areas are of this nature. Yuan (2001) has described *field objects*, or objects with continuously varying internal structure. Time adds a new dimension to this discussion, because of the many types of temporal change (Peuquet 2002), as does the third spatial dimension.

The second problem concerns the earliest stages in the acquisition of geographic knowledge. Both relational and object-oriented approaches are based on tables, and imply the existence of well-defined sets of objects and attributes. But although this model may fit well to mature mapping processes, it leaves much to be desired as a representation of geographic exploration, and the processes by which human observers build conceptual understandings of the world around them. Classification is often regarded as the first stage of any scientific analysis, but there are certainly stages of observation that occur well before the establishment of classification schemes. And although maps identify *features* on the earth's surface, the identification of features represents a fairly mature level of understanding of a landscape, and a high degree of consensus.

Consider, for example, the explorations of Lewis and Clark, or the fieldwork of Alexander von Humboldt or Charles Darwin. All of these explorers made extensive records of their travels and observations, but made little use of the tables that dominate in geographic data modeling. Instead, one might characterize their observations as largely unrelated and uncoordinated notes of the form $<x,z>$, indicating that at some location in space–time x a property z was observed. Only later was it possible to assemble these unrelated observations into tables, and to conduct the kinds of analysis now associated with GIS. Goodchild *et al.* (2007) have termed this the *atomic* form of geographic information, and have shown how discrete objects and continuous fields can be conceptualized from many such atoms, along with object fields, field objects, and the tables of high-level geographic data modeling.

Conclusion

Flat files, relational models, and object-oriented models represent three stages in the evolution of geographic data modeling. Each approach is more comprehensive than the one it replaces, and although GIS developed initially based on the realization that several types of mapped data could be represented using the same basic approach, today's object-oriented data models encompass a much wider range of geographic data types that extend far beyond the traditional domain of cartography. Each approach has emerged from the mainstream computing industry, as computers became more powerful and as the technology of database management became more sophisticated; and each approach has been quickly adopted and adapted to the needs of geographic information.

No single approach can accommodate all data types, and the number of such types continues to increase as geographers and others explore the modeling of complex, dynamic phenomena. In each of the three stages pressure has built to accommodate a wider range, and work-arounds and extensions have been added, increasing the complexity of the approach and reducing its essential coherence. Finally a new solution has emerged from the mainstream, and has been adopted with enthusiasm, restarting the cycle.

The final section argued that the current emphasis on object-oriented designs is inadequate in two respects. It fails to model continuous fields in appropriate ways, and to prevent users from confusing data sets based in continuous-field and discrete-object conceptualizations. Moreover it fails to accommodate the earliest stages of field observation, leaving a large and important phase of scientific research without effective formalisms and computational support.

Whether the cycle will begin again remains to be seen. Previous cycles have been initiated by developments in the mainstream computing industry, and there are no indications that mature technologies are about to emerge to solve either of these problems. Moreover the GIS software industry is driven by its largest commercial customers, many of whom operate in areas such as asset management where discrete-object conceptualizations are more appropriate. Nevertheless the inability to handle continuous fields effectively is a cause of substantial confusion in GIS applications, and significantly increases the difficulties of learning GIS. Perhaps the use of encapsulated methods, as proposed in this paper, will provide a short-term solution.

The geographic world is infinitely complex, and its useful and accurate representation in the limited and discrete space of a digital computer remains a challenging problem, just as earlier the need to represent the world using pen and paper had its own severe limitations. The formal nature of a database inevitably favors certain types of geographic knowledge over others, and tends to work against knowledge that is subjective, inconsistent, and otherwise at variance with the principles of scientific measurement. Much progress has been made in the four decades since the first GIS experiments, but much important research and development remains.

Note

1   The support of the U.S. National Science Foundation through Award BCS 0417131 is gratefully acknowledged.

References

Arctur, D. and Zeiler, M. (2004) *Designing Geodatabases: Case Studies in GIS Data Modeling*, Redlands, CA: ESRI Press.

Cova, T.J. and Goodchild, M.F. (2002) 'Extending geographical representation to include fields of spatial objects', *International Journal of Geographical Information Science*, 16(6): 509–532.

DeMers, M.N. (1997) *Fundamentals of Geographic Information Systems*, New York: Wiley.

Foresman, T.W. (1998) *The History of Geographic Information Systems: Perspectives from the Pioneers*, Upper Saddle River, NJ: Prentice Hall.

Goodchild, M.F. (1988) 'Stepping over the line: technological constraints and the new cartography', *American Cartographer*, 15: 311–319.

Goodchild, M.F., Yuan, M. and Cova, T.J. (2007) 'Towards a general theory of geographic representation in GIS', *International Journal of Geographical Information Science*, 21(3): 239–260.

Longley, P.A., Goodchild, M.F., Maguire, D.J. and Rhind, D.W. (2005) *Geographic Information Systems and Science*, 2nd edition, New York: Wiley.

Mark, D.M. and Csillag, F. (1989) 'The nature of boundaries on "area-class" maps', *Cartographica*, 26: 65–77.

Peuquet, D.J. (2002) *Representations of Space and Time*, New York: Guilford Press.

Stehman, S. (1996) 'Estimating the kappa coefficient and its variance under stratified random sampling', *Photogrammetric Engineering and Remote Sensing*, 62: 401–407.

Yuan, M. (2001) 'Representing complex geographic phenomena with both object- and field-like properties', *Cartography and Geographic Information Science*, 28: 83–96.

Zeiler, M. (1999) *Modeling Our World: The ESRI Guide to Geodatabase Design*, Redlands, CA: ESRI Press.

Zhang, J.-X. and Goodchild, M.F. (2002) *Uncertainty in Geographical Information*, New York: Taylor & Francis.

6 *theirwork*

The development of sustainable mapping

*Dominica Williamson and
Emmet Connolly*

Introduction

theirwork is a participatory online 'open' mapping project put together by project participants' local knowledge and direct experiences of their lived environment with the aim of creating a democratic and first-hand, local definition of place. It rejects proprietary mapping software, generally characterized by copyrighted and prescribed visualizations of spaces. By opposing these authoritative, top-down systems of classification that are disempowering and homogenizing the world we live in, *theirwork* opens up the possibility of creating an emancipatory, continuously evolving mapping that is situated in a given space and addresses its creators' – which are simultaneously also its potential end-users – main concerns. Thus the map becomes rooted in local identity, fulfilling the needs and reflecting the interests of the community. The custom-made, open-source software and approach encourages a more reflective reading, viewing and understanding of one's environment and facilitates the recording as well as the protection of traditional knowledge and communal experience of space, which ultimately can then also be shared between different communities.

theirwork in the local mapping context

Traditionally maps have acted as a form of literal and abstract representation. The standard map is a precise top-down cartographic representation of a geographic terrain, a visualization of place. Fundamentally, maps are used to provide a view of data that is manageable for particular groups and uses. They are designed to be easily understood and represent selected information that is scaled down for ease of use. Similarly, a looser definition of cartography may allow us to consider that any form of data abstraction or representation based on a location is a map. This framing has seen maps being used as a medium to communicate ideas beyond the scope of physical geography, such as the amount of pollution in a given neighbourhood. More recently, the environmental movement has adopted mapping as a form of communication. Green Map System (1999), an organization based in America, in particular, formed around the notion of almost exclusively using maps to further the cause of environmentalism.

Green Map System encourages communities to gather data about their green facilities and spaces. Communities are encouraged to map toxic hotspots, good places to view stars, or green businesses for instance. The end result of this community process-orientated mapping is that Green Maps typically present a variety of ecology-related points of interest on a map.

Similarly, Parish Maps is led by a British environmental arts organization called Common Ground. The project calls for a communal mapping of villages, towns and cities. Here people participate in mapping what concerns them about their place in order to protect their local distinctiveness (Common Ground 1983). The boundaries of the map are determined from the outset by the Parish boundaries limiting the project to producing static maps of UK villages (Crouch and Matless 1996: 237–9). Whereas Parish Maps set up boundaries by use of their terminology and by often veering towards a rosy-type Ordnance Survey representation of place (Crouch and Matless 1996), Green Maps set boundaries in terms of what type of data constitutes 'green' data. Both favour quantitative data to the exclusion of qualitative data and have a tendency to freeze information by restricting people from adding to the picture.

A plethora of other local mapping initiatives have come about, such as indigenous communities fusing their traditional mapmaking techniques with other mapping processes to fight for their rights – and with success (Harrington 1999: 2). These call for democratic mapping processes and have attempted to reframe who and what a map is for. Indigenous maps are often made, used, re-made and used again in a communal setting. Both indigenous and standard maps can become powerful dynamic educational and decision-making tools (Common Ground Project 2008).

Participatory geographical information system projects have burgeoned recently and focus on ensuring the voice, and so the map, of 'the other' is heard/seen (see Cope 2008; Elwood 2008; Ghose 2001; Kitchin 2002). Such projects gather and input data with the community using GIS software. Data sets can then be modelled geographically to raise community issues to influence policy-making. Qualitative data is also being gathered in this way to ensure that issues are not excluded from the map and so that sophisticated data sets can represent realistic views of how people are operating in a place (Kwan 2007: 175). Modern technology is set to revolutionize the production and distribution of maps further than this. The global Web has made the means of production available to almost everyone with access to it. Wikipedia is a prime example of an evolving knowledge resource based on online community data editing, while Google Maps allow the creation of custom-made maps, substantially lowering technical difficulties of map creation.

All the above examples use participatory mapping methodologies but some reject proprietary mapping software. For instance, Google Maps is founded on the open source software approach, allowing free access, adaptation and re-distribution of software without any or few copyright restrictions. This free and open access to all must be seen as a much more sustainable, holistic approach to mapmaking. Within the field of Google Maps this open, integrated, community-led approach to developing a project has much to offer, in particular to groups who wish to mobilize a community wanting to feel invested in a movement, as is the case with Green Map System.

theirwork: *the project*

theirwork is an online open map. Open source software that drives the map is available for anyone to use or re-appropriate, rejecting a proprietary approach. Loe Pool in Cornwall, Britain (the county's largest natural lake) is the first area to be mapped by the software. While mapmaking is at the centre of the project and is used to ground the collected data, it is also used to root the project in real-time space. *theirwork* works closely with end-users, who are treated as co-developers by walking, talking and recording in its landscape. The mapmaking it seeks to produce is grounded in multiple perspectives; therefore multiple voices and autonomous experiences are documented via first person sensory experience and through a community's felt experience of landscape. The project is open, inclusive and non-hierarchical in both form and content. The software (form) and data collection (content) are symbiotic and mutually supportive in terms of 'openness'. *theirwork* software rejects a top-down system of classification or taxonomy and adopts instead a system of crowd-sourced labelling, or what has been dubbed folksonomy. Regarding authoritative and hierarchical taxonomic systems as disempowering, the folksonomic approach enables the *theirwork* participant, who works online, to collaboratively generate open-ended labels for mapped data.

Using open methodological frameworks, *theirwork* ensures that the development, production and dissemination of local definitions of place are gathered and visualized through soft (qualitative) and hard (quantitative) data collection, without any restriction on re-use. Importantly, such innovations guarantee that local definitions of a place are presented using sustaining, rejuvenating software. Foregoing other top-down systems that often produce hegemonic systems and organizations (such as copyrighted Ordnance Survey maps and copyrighted Geographical Information Systems (GIS) data), *theirwork* innovates and builds upon the movement called Green Mapmaking.

Critiquing green mapmaking

theirwork adopts three cognate disciplines: psychophysical geography,[1] phenomenology[2] and ethnography.[3] These complementary approaches have created a methodological framework, through which open data are sourced and collected. Ethnographic methodology ensures multiple voices construct the map: the phenomenological approach ensures autonomous experiences are documented via first person sensory experience, and through a community's felt experience of landscape. Last, a psychophysical geographic approach ensures the map is emotive and deeply personal. All three approaches ensure the map is grounded in locality, subjectivity and a lived experience of place. A common discourse exists among cognate genres committed to plurality, locality and subjective interrelations of body-landscape. Immediately, the terms Green Mapmaking, indigenous mapmaking and bioregional mapping come to mind, each advancing and augmenting current mapmaking praxis. *theirwork* is situated within all the above, but seeks to advance the area of soft data collection and challenges existing software that is used by many mapmakers – as although existing maps often work on the principle of open content and sharing, many use closed systems of software and licensing production to make their maps.[4]

In line with bioregional mapmaking and the writings of Ben Whelan (2002: 36), *theirwork* calls 'the community into the process of mapmaking' where 'the charted landscape is filled with the stories of its dwellers and an intimate knowledge of their ecosystem'. Whelan's (2002: 36) call, radical and compassionate, seeks to deepen 'the communion between human and nature' and create maps 'that can accommodate multiple levels of reality'.

Bioregional mapmaking's allegiance to the non-human world grew directly out of various genres of indigenous mapmaking – all wayfinders deeply connected to the landscape. Mapmaking that calls for a human appreciation and protection of the landscape and its indigenous species, has in turn created a genre of urban bioregional mapmaking called Green Mapmaking. *theirwork* is situated within Whelan's inclusive discourse of Green Mapmaking praxis; for example, *theirwork* 'seeks to energize local knowledge and mobilize citizens into action' in order to address 'greenness' (Green Map System 2007).

Green Mapmaking at first glance appears an inclusive term, because it allows communities to shape their own picture of the present and future, by supplying toolkits that encourage them to chart their natural and cultural environment. These toolkits centre on a set of global Green Map Icons that the community must use in order to label their project a Green Map. However, this model still operates through a structure of exclusivity. Although these toolkits are a marked improvement upon Ordnance Survey maps and other traditional mapmaking systems, they are still partly exclusive in terms of creation, access and usage – in short, they restrict innovation. First, the structural and visual boundaries of Green Maps are often defined by criteria, which in turn are usually defined by a steering committee. Second, software is difficult to use when a community want to reproduce icons digitally. In an online environment, due to copyright restrictions, icons are difficult to use. Third, data is hard and lacks qualitative insight. Definitions of a locality tend to be shaped by hard data collection only, because data is often fitted into this icon set. Last, icons, although a powerful visualization tool, are aligned in our cultural memory with traditional topographical maps and their boundaries.

Green Mapmaking is situated within the wider problematic discourse of sustainability. For example, *theirwork* was born out of a concern for the environment, fuelled and shaped by an escalating political rhetoric that centres on the concept of sustainability. The European government is attempting to translate and implement sustainability through a practice-based legislative process, whereby industry is forced to comply with greening initiatives (EurActiv 2009) and general 'lay audiences' are targeted by local government bodies to construct social well-being and encourage new communities via new initiatives (United Nations 2008).

The concept of a Green Map was created in New York over a decade ago. At first, Green Mapmakers did not directly use sustainability and its associated terminology. However, bioregional mapmakers and Green Map System started to use the word to situate their work within a wider socio-context. 'The impetus for creating and teaching these new skills of sustainability [and mapping] are coming from residents in scores of places who refuse to see their social and ecological capital either under-utilized or squandered' (Harrington 1999:6).

Here in the UK, different fields of knowledge work to gain funds that will help them address the social, environmental or economic aspects of sustainability, but few agree about what the concept means in its entirety and even fewer are able to implement it in practice:

> [. . .] problems arise in part because the sustainability of the human enterprise in the broadest sense depends on technological, economic, political, and cultural factors as well as on environmental ones and in part because practitioners in the different relevant fields see different parts of the picture, typically think in terms of different time scales, and often use the same words to mean different things.
>
> (Daily *et al*. 1995: no pagination)

theirwork recognizes the confusion and disparity that surrounds the word sustainability. Most importantly, *theirwork* believes the term sustainability exists and operates within a number of governmental hegemonic discourses, i.e. the term itself is continually produced within legislative power structures. For example, Agenda21 officers were situated in each UK district council by the late 1990s. Their job was to help find sustainable solutions to problems within their local community.[5] In contrast, *theirwork* does not centre mapmaking praxis on generic or legislative definitions of sustainability, but rather encourages dialogue that supports the re-formation of self, community and place. *theirwork* does not seek to overturn generic understandings of sustainability, but rather seeks a more complex understanding and proliferation of the term via local 'grounded' definitions. *theirwork* therefore builds on Green Mapmaking and sustainable discourses, but has created innovative strategies within the genre of bioregional mapping, particularly in the following areas: mapping software, online access and the gathering of soft data.

Possible solutions, coming from open source

Having identified fundamental problems and restrictions inherent in existing models of Green Mapmaking, the question of how to define an alternative framework presents itself. The flexibility provided by Internet mapping has already been explained, and in the case of *theirwork* was seen as the most likely medium to allow the type of open-ended activities that traditional Green Maps cannot.

There is obviously an established body of work in the field of Web-based mapmaking that requires critical appraisal. First, however, it is necessary to survey the wider terrain of computing and take stock of what influences can be garnered from its politics and philosophies. An immediate parallel can be drawn between the wider green movement, from which Green Mapmaking emerged, and the open source software movement. In an effort to establish a more holistic and sustainable approach to mapmaking in general, it was deemed necessary to focus on each of the constituent parts that the framework takes, and ensure that the approach is consistent and self-propagating. Hence, a focus on the ideologies of software development was central to the maturation of *theirwork* as a coherent movement.

The open source movement at its core stands for the development of source code (the algorithms and computer logic written by computer programmers to create software) in a completely open and free way. Pragmatically, this manifests itself as a methodology of making code freely available to anyone who may wish to access it for any purpose, unconditionally. Concurrently, open source is for many a philosophical approach to software development, and is seen as the only truly sustainable approach to software development. Open source code may be shared, studied, copied, reused, modified, built upon and redistributed in any way. As such this model has made possible innumerable software projects that would otherwise have been almost impossible to realize (the most popular examples include the Linux operating system and the Firefox Web browser, both used by millions of computer users).

The possibilities of the model are highlighted by open source evangelist Eric S. Raymond in his seminal 1997 essay *The Cathedral and the Bazaar*, in which he compares the development approach to 'a great babbling bazaar of differing agendas and approaches', all of which create a finished product that could never have been designed or executed by a single architect.

In today's world of corporate global software giants, whose billions are based upon the materiality and inaccessibility of code, this can seem to be a revolutionary set of ideas. Yet in both its execution as a model for making possible new forms of collaborative work, and its philosophical underpinnings of sustainability and openness, it is an essential component in and influence upon a computer-based mapping solution.

In the earliest planning stages of the project, it was resolved that in order to improve upon the existing framework of Green Mapmaking, the entire back-to-front process of it should be executed in line with the ideals of the project as a whole. The ongoing development of the tool, the process of creation, was taken to be as important as the final artefact, and consistent with the ideology that drove its inception. Thus, although the use of open source software was in line with the spirit of *theirwork*, this alone was not sufficient to constitute a holistic approach. At every stage of development, decisions were consciously influenced by the desire to create a project that would at every turn reinforce itself. As this concept developed, the approach came to be labelled 'sustainable software', and drew together influencing characteristics from a number of disparate fields, combining select strands from each into what was hoped to be a coherent whole.

The architectural method of adaptive design, that of designing and building to ensure that a system retains enough inherent flexibility to be modified (or even encourage modification) that had not originally been considered, was an influence on the planning of the project outcome. The Slow Food movement, which encourages a change of pace and even lifestyle in order to reassess priorities and values, was another.[6] Apart from open source licensing, the object-oriented approach to writing computer code, which ensures that each part of the code is modular and easily replaceable, was an influence from the arena of technology. The copyleft and Creative Commons (2008) movements that eschew the traditional concepts of information ownership in favour of a more liberal approach to content sharing (of which more later) were studied carefully. Some of these ideas were adapted quite literally, but were also taken as philosophical or political approaches, helping to shape the concept of sustainable software.

With a driving ideology defined (or as clearly defined as any set of ideas which have at their core the intention to be as flexible as possible), the question of how to actually implement the project naturally arose. It was decided at an early stage to make the software Web-based to allow for a process of rapid development and iteration and allow a maximum number of potential participants. Another, more pragmatic, reason was to facilitate the fact that the two main contributors to the project live in different countries; almost all communication was carried out via a combination of email, phone and instant messaging. Likewise, the development of the project was largely carried out 'in the open', with participants contributing via the *theirwork* blog, wiki and online forum (Figure 6.1).

For reasons that should be obvious from the influence of open source, it was decided not to pay for the right to use commercial mapping software. One of the next obvious approaches when creating Web-based maps is to use an already-available service, such as Google Maps. It is relatively simple to create what is known as a Google Maps mashup; that is, taking an existing

Figure 6.1 Participants' contribution via the *theirwork* interface. Source: author
screenshot.

map, and overlaying one's own data on to that map. As an immediate technical
solution, a Google Maps mashup would appear to be the easiest option.[7]
However, close inspection of Google's terms and conditions revealed that
the licensing it bore did not meet the strict guidelines that had already been
established in relation to software licensing for the project. Nor did any
existing open source mapping toolkit meet the needs of the project. It was
eventually decided to build a custom software solution, and make it available
to the public as open source software. It should be noted, however, that a
number of existing open source toolkits were used to create smaller parts of
the tool, combining to create a new whole. Without the ability to reuse and
adapt the code that already existed within the ecosystem of the open source
community, it would have been practically unfeasible to develop such a
complex system.

Creating a base image for the map (i.e. a top-down view of the lake where
the project was piloted, on which to plot the data) was not an easy process.
Again, licensing restrictions proved a point of contention; the now-
controversial laws surrounding Ordnance Survey data meant that purchasing
the map data for the lake was ideologically and financially out of the question.
Although there are nascent communities such as OpenStreetMap (2004)
currently endeavouring to make geodata freely available in the UK,[8] no efforts
existed in the geographic location that *theirwork* focused on. There was no

pre-existing, freely available data on which to build. In order to obtain the data, GPS units were used to record the track points of a walk around the perimeter of the lake, and specific points of interest were marked along the way. The result was a matrix of latitude and longitude GPS coordinates, which were then loaded onto a computer, where pre-existing open source software was used to generate a simple line drawing of the lake's outline. This outline was then annotated by hand to create a defined background map on which data points could be plotted. This was a laborious and technical process, and represents one of the major remaining obstacles to the breakthrough and popularization of people-powered mapping; it will inevitably be overcome by the proliferation of user-friendly convergent hardware that integrates GPS with popular consumer recording devices, such as cameras.

The fully developed beta version of the software consists of a Web-based Google Maps-like interface, by which the user can interact with a map of the lake. A number of data points that have already been added by other users are overlaid on the map and may be clicked for more information. What makes *theirwork* slightly different from other mapping software is the ability for users to immediately add their own points of interest to the map directly at any time. They may also edit existing points to improve them as they see fit. This open model of community data editing is taken directly from the wiki model (the best known example of which is the online encyclopaedia Wikipedia), in which participants may add or edit any page on the website. This distributed model of content creation can work remarkably well in some cases, and is surprisingly capable of 'self-healing' in cases of vandalism, whereby a subsequent user notices and immediately rectifies an existing error.

At the same time, a completely open data system such as this could make for a chaotic set of data, if not presented in a logical manner. The question arises: what is a sustainable model of group data classification? Green Maps have encountered the problem that their maps can be too narrow in subject if a strong editorial control is exerted, and too chaotic and unstructured if free rein is permitted.[9] How can people be empowered to add whatever type of data they wish to the map, but also have a coherent picture emerging from the map as a whole?

Fortunately, computers are adept at taking a lot of information and shuffling it, or slicing and dicing it, in any way. Many websites with user-generated content have experienced a similar problem recently, attempting to classify an open data set without imposing structure. The aim is to somehow capture (to paraphrase a book title on this topic) the Wisdom of Crowds (Surowiecki 2004), and allow an emergent picture to develop from the teeming mass of individual actions happening within a system. The solution here is to reject a top-down system of classification, or taxonomy, and adopt instead a system of labelling, or what has been dubbed folksonomy. This involves rejecting any notions of hierarchical classification, and allowing users to tag their data with keywords that describe it instead. A data point has many keywords pinned on to it, instead of being placed into a single category. This actually opens up the process considerably, and leads to a much more creative way of adding data. Users now have the freedom to use the map in ways that the map designers may never have even conceived. The map becomes an adaptive, open-ended, and sustainable ecosystem of data.

At the data output stage, when trying to discover or extract all of the data that a user is interested in, they do not dig down into a category to find the relevant items, but rather filter out all items by keyword. This may be thought of as viewing a cross-section or slice of all data, except that even within this single slice, there exists a lot more information still to be mined; many more strata of keywords that may line up, or move off in a different direction. The whole experience makes for a much richer data process. This approach works well for open data in mapping, as it means that we can dismiss concerns about misclassified information, or editorial control, and concentrate on extracting a meaningful signal from the rich information set. This opens up a route for an entirely new type of emergent, community-developed map creation that coherently represents the combined impressions of an unrelated group of self-interested actors, and conveys a truly distributed simulation of a geographic space.

Developing open data, out of place-based mapping

> The voice of the participant, rather than the voice of the researcher, will be heard best when participants not only provide the data to be analysed, but when they also contribute to the questions that frame the research and contribute to the way data are analysed.
>
> (Ezzy 2002: 64)

The above quote encapsulates why an ethnographic approach was necessary in this project. In terms of mapmaking praxis, construction of the map has been an entirely de-centred process and authoritative models of data collection and transcription have been overturned. The application of ethnographic methodology ensures multiple voices construct the map. Within this work the relationship of emotion, memory, and sensory engagement with the landscape was mapped. First, data was sourced while walking, talking and recording with participants on the landscape. After an initial recruitment period and focus session, each co-developer chose a location for a 'one-to-one' walk that in some way was connected to the lake. Co-developers chose the date and time – some brought their binoculars or dog along, others even brought 'somebody else along'. The co-developers were helped in tracking the walk; sites of interest, objects, plants and animals, favourite places, memory spots and stories connected to the place. Places were noted using cameras, notebooks, a GPS unit and a dictaphone. A framework of open and closed questions was asked. Answers to open questions, such as 'What do you feel about the lake?' were geo-tagged. The 'type of walk' (their special walk) became an integral part of data collection and data analysis. These processes helped capture the walk and created a supplementary resource to each recorded conversation that took place, which was then transcribed.

In the spirit of ethnographic methodology, transcription and coding of data was a mutually inclusive activity (all information was verified with co-developers). Some of the codes that developed from the walks were words such as rocks, water, agriculture, birds, meditation, trees, fields, memories, fish and events. In a paper-based workshop, co-developers then jointly discussed the codes and each shared their record of the walk from memory. Memories were added to the discussed codes. Importantly here, qualitative data became coded by the co-developers and not by some distant and 'removed' ethnographer. To this end, paper-based tags were ready-made for the map interface. A sort of starter kit had been created, effectively introducing co-developers to tagging or folksonomy. Qualitative coding methodologies in turn introduced the community to the art of good folksonomy. This is an important issue, because it deepened the practice of folksonomy and helped to reflect on it in a practical and academic manner.

A computer workshop then tested the beta version of the sustainable software. Each co-developer put marks on the map, using latitude and longitude figures supplied from the archive of walks and paper-based workshop. They tagged their marks efficiently and with ease, having been introduced to the concept of folksonomy in the paper workshop. When things started taking shape onscreen the mood in the workshop room was electrifying. Everyone watched their places appear on the map – and all the efforts and concepts that must at times have seemed utterly puzzling started to make sense and finally paid off. Technical problems were fixed as and when they arose. Co-developers' views, feelings and ideas for the future of the software, as well as ideas for new data, were taken into account.

Outdoor events have since become interchangeable with ongoing paper and computer workshops. All types of place-based mapping happenings have been called for by the co-developers, and are enabling the gathering of data that was not pre-determined. For instance, moth migration nights, stargazing gatherings, butterfly balls, drawing picnics and plastic bag counts have taken place. The geo-coded data is challenging how the base map could look and function, and is drawing in experts in the field of flora and fauna and qualitative research.

At present, when data is added, co-developers either leave trails of red dots where they have been recording a walk or they add to pools of information where groups have gathered. For instance, a moth and bat night focused on three spots at the mouth of the lake, and became like three micro-maps of fascinating creatures and facts (Figure 6.2). These red marks could become a sea of pictures, telling a tale of moths in this area. At the moment pictures are uploaded to a separate space, to a group Flickr (2004) account. As funding is applied to develop the project further, the co-developers will become involved in the application, asserting what they think should be

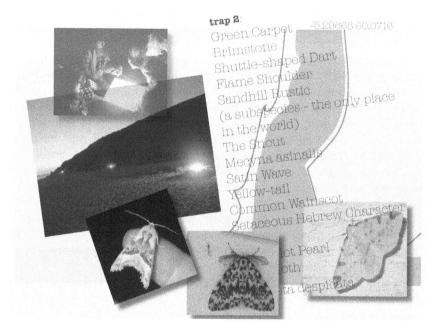

Figure 6.2 Moths from left to right: Angle Shades trap 3, Black Arches trap 1, Brimstone trap 1 & 2. Photographs by co-developer Nikki Schneider. Source: authors.

developed next. Their priorities so far are: to make the base map more sophisticated; to make their map entries about each place more editable; to style the forum and to be able to guest blog (at the moment the project initiators and a few others are blogging).

theirwork in summary

theirwork's development philosophy and production activity seeks to offer an example of sustainable mapping in practice. Open-ended in nature, the project seeks to help communities to care for a place through the creation of a shared language through open and unrestricted content. As *theirwork* is in its infancy, it is as yet impossible to make an objective assessment of its effect on software development and mapmaking at a bioregional level. At a subjective level however, memorable place-based mapmaking experiences are taking place because of this type of mapping, and are affecting how a small group of people view an area they regularly spend time in. Due to *theirwork* activities they are seeing things they never saw before, learning things they did not know before and importantly, are collectively finding a way to share knowledge about a place. It appears that encouraging the idea of effective and active sustainability productivity is attached to place-based mapping. *theirwork* will therefore continue to support this angle of inquiry and is inspired by theorists such as David Abram (1996: 273–4), who calls for re-habitation through action of place and of body, so that we can inhabit places like coastal forests and grassland again; and writers such as Jay Griffiths (2006: 16), who are against the closing down and patenting of open knowledge: 'Not for nothing is wild knowledge called "common knowledge." Common knowledge is free, open, unenclosed – and "free" financially: it must not be bought or sold for profit'.

Akin to public participatory geographical information system (PPGIS) projects, this project then calls on a political level for everyone to have access to spatial data and to be able to add to the picture, so as to help develop and protect 'traditional knowledge and wisdom from external exploitation' (PPGIS 2008: no pagination). Second, it demonstrates that if carried out appropriately, participatory mappings are political and powerful ways of learning and sharing how to use a place (Kitchin 2002: 57). Third, as Mei-Po Kwan (2007: 171–2, 175) advocates, the use of qualitative and quantitative geographical research enables a more informed and so realistic set of data. Crucially, however, the project advances local mapmaking by rejecting complex and expensive proprietary software. It turns to online mapmaking as a way to create an open, inexpensive and accessible way of building data and of proffering an open Green Map. Although projects that utilize GIS software and build qualitative data with communities are advancing areas in the field of participatory mapping, they are also struggling due to GIS data restrictions (Elwood 2008: 73–5, 81–3; Ghose 2001: 142–4, 156–8). This software can build complex visualizations of data, and when used for the right occasions, places, peoples and budgets can build powerful results and case studies (for example, Kitchin 2002; Kwan 2007). Such projects argue for a spatial data for all, and work at grass-root level, but more projects could take place if other forms of software, which are less complex and expensive, were available. *theirwork* does not claim to be able to replace such software but rather hopes to demonstrate that other software routes to mapping can deliver mixed methods approaches.

Combining psychophysical, phenomenological and ethnographical strategies is vital to local mapmaking. Qualitative researcher Iain Edgar (2004) encourages a visual ethnography and has developed a methodology that is not just of external images but is of an imagination of images from within. Such approaches applied to phenomenological mapmaking could empower communities in terms of getting them to visualize, claim and know their place. Continuing to combine hybrid ethnographical field research with computer workshops, which utilize a folksonomic approach to coding data, could continue to enable different types of people to engage with a place.

theirwork will continue to build on folksonomic developments to help assess how users find, add and extract data. For instance, it will look to see if data is tagged using time, as well as imagination and hard fact. Like technologists Joe Lamantia (2006) and Pietro Speroni di Fenizio (2005), *theirwork* feels that the right use of this labelling system will allow us to 'see changes in the culture we are living in' (di Fenizio 2005: no pagination). In the future *theirwork* will concentrate on how it can continue to adapt the software and present its data in different ways. As technologist Matt Biddulph (2006) observes, anyone should always be able to get data out of the format it is in because no one ever knows when they may need to re-purpose it.

Accordingly, *theirwork* continues to embrace a digital psychogeography within this new neogeographical realm, and believes that such political, ethnographical and technological mapmaking will enable different types of communities to share and exchange data, information and knowledge.

Notes

1 Psychophysics is the 'branch of psychology that deals with the relationships between physical stimuli and sensory response' (The Free Dictionary, www.thefreedictionary.com/psychophysics). *theirwork*, inspired by such books as *Textures of Place* (Adams *et al.* 2001) relates the psychophysical to geography.
2 Phenomenology is harnessed by *theirwork* in its broadest sense, 'addressing the meaning things have in our experience, notably, the significance of objects, events, tools, the flow of time, the self, and others, as these things arise and are experienced in our "life-world"'(Stanford Encyclopedia of Philosophy, http://plato.stanford.edu/entries/phenomenology/).
3 Ethnography involves intensive fieldwork and data gathering and 'may be pursued in a variety of settings that allow for direct observation of the activities of the group being studied' (Moustakas 1994: 1–2).
4 The *Green Map* network is based in New York. It uses open source tools but its own iconography is copyrighted, although the wider *Green Map* community, and not just the head office have developed the Green Map Icons.
5 For example, the Royal Borough of Kingston upon Thames' Local Agenda 21 implementation began in 1988, see www.kingston.gov.uk/environment/agenda21.htm.
6 It is in the field of eco-design, due to its prototyping practice, and its embracement of the notion of 'slow' (by Manzini, Fuad-Luke and *slowLab* (2005)) that *theirwork* looks toward the Slow Food movement.
7 Aside from the start of Open Green Map and a Google Green Map-mashup (Dear Green Place 2006 whose code can be traced back to Emmet Connolly's Galway Free WiFi hotspots map (http://thoughtwax.com/sandbox/galway-wifi), few Green Maps have been geo-coded. If a Green Map has been geo-coded in the past, it has usually happened because a project has been linked or initiated by a more-than-profit organization or a learning institute that has access to GIS tools.
8 OpenStreetmap is a mapping project that became exasperated by the restrictions Ordnance Survey were placing on more-than-profit organizations and individuals. Lauded and used by like-minded activists such as the Free Our Data leader Michael Cross, it helps people through workshops and outdoor activities to make copyright-free base maps.
9 For instance, at a Green Map conference in Bellagio, Italy (a week-long event sponsored by the Rockefeller Foundation in 2002) criteria debates ensued following the presentation of a Green Map that had a McDonalds restaurant placed on it.

References

Abram, D. (1996) *The Spell of the Sensuous*, New York: Vintage.

Adams, C., Hoelscher, S. and Till, K. (2001) *Textures of Place: Exploring Humanist Geographies*, Minneapolis, MN: University of Minnesota.

Biddulph, M. (2006) 'Open data movement: The next wave of open source', presented at *Helsinki Institute for Information Technology*, 17 August. <http://jyri.blip.tv/file/62508>.

Common Ground (1983) website, <www.commonground.org.uk>.

Common Ground Project (2008) *Common Ground Community Mapping Project*, <www.commongroundproject.ca/projects>.

Cope, M. (2008) 'Becoming a scholar-advocate: Participatory research with children', *Antipode*, 40(3): 428–35.

Creative Commons (2008) website, <http://creativecommons.org>.

Crouch, D. and Matless, D. (1996) 'Refiguring geography: Parish maps of common ground', *Transactions of the Institute of British Geographers*, 21(1): 236–55.

Daily, G.C., Ehrlich, P.R. and Holdren, J.P. (1995) *The Meaning of Sustainability: Biogeophysical Aspects*, Washington, DC: United Nations University and World Bank, <http://dieoff.org/page113.htm>.

di Fenizio, P.S. (2005) *Tagclouds and Cultural Changes*, <http://blog.pietrosperoni.it/2005/05/28/tagclouds-and-cultural-changes>.

Edgar, I. (2004) 'Image in ethnographic research', in A.I. Alfonso, L. Kurti and S. Pink (eds) *Working Images: Visual Research and Representation in Ethnography*, London: Routledge.

Elwood, S. (2008) 'Grassroots groups as stakeholders in spatial data infrastructure: Challenges and opportunities for local data development and sharing', *International Journal of Geographical Information Science*, 22(1): 71–90.

EurActiv (2009) *Sustainable Development: EU Strategy*, <www.euractiv.com/en//sustainability/article-117544>.

Ezzy, D. (2002) *Qualitative Analysis: Practice and Innovation*, London: Routledge.

Flickr (2004) website, <www.flickr.com/groups/loepool/>.

Ghose, R. (2001) 'Use of information technology for community empowerment: Transforming geographic information systems into community information systems', *Transactions in GIS*, 5(2): 141–63.

Green Map System (1999) website, <www.greenmap.org>.

Green Map System (2007) *Think Global, Map Local!* <www.greenmap.org/greenhouse/en/about>.

Griffiths, J. (2006) *Wild*, London: Hamish Hamilton.

Harrington, S. (1999) *Giving the Land a Voice: Mapping Our Home Places*, 2nd edn, Canada: Salt Spring Island Community Service Society.

Kitchin, R. (2002) 'Participatory mapping of disabled access', *Cartographic Perspectives*, 42: 50–60.

Kwan, M. (2007) 'Hybrid GIS and cultural economic geography', in A. Tickell, E. Sheppard, J.A. Peck and T. Barnes (eds) *Politics and Practice in Economic Geography*, London: Sage.

Lamantia, J. (2006) *Second Generation Tag Clouds*, <www.joelamantia.com/blog>.

Moustakas, C. (1994) *Phenomenological Research Methods*, London: Sage.

OpenStreetMap (2004) website, <www.openstreetmap.org>.

PPGIS (2008) *About PPGIS Website*, <www.iapad.org/ppgis/ppgis.htm>.

Raymond, E.S. (1997) *The Cathedral and the Bazaar*, <http://catb.org/~esr/writings/cathedral-bazaar>.

slowLab (2005) website, <www.slowlab.net/index.html>.

Surowiecki, J. (2004) *The Wisdom of Crowds*, New York: Random House.

United Nations Division for Sustainable Development (2008) website, <www.un.org/esa/sustdev/documents/agenda21/index.htm>.

Whelan, B. (2002) 'Mapping reality', *Convergence Magazine*, 36–7, <www.sustainable.ie/convergence/magazine/convergence2002.htm>.

7 Cartographic representation and the construction of lived worlds

Understanding cartographic practice as embodied knowledge

Amy D. Propen[1]

Introduction

> Cartographic practice inherently is learning to make projections that shape worlds in particular ways for various purposes. Each projection produces and implies specific sorts of perspective.
>
> (Haraway 1997: 132)

Traditionally, the field of cartography has been rooted in an objective, standardized set of practices that purport to convey accurate and correct models for ways of knowing. Even prior to the advent of the spatial sciences, it has been the goal of cartographic practice to:

> produce a "correct" relational model of the terrain. ... Similarly, the primary effect of the scientific rules was to create a "standard" – a successful version of "normal science" – that enabled cartographers to build a wall around the citadel of the "true" map.
>
> (Harley 1992: 234–5).

This notion of the true, correct, objective map is one that leaves little room for speculation regarding the models of terrain it produces, as well as the more far-reaching implications of what such speculation might allow for. Without accounting for or allowing for subjective levels of analysis, the fields of cartography and GIS are thus relinquished from any level of responsibility or accountability for the knowledge claims implicit in their work. With this caution in mind, I argue instead for an engaged and responsible approach that understands cartographic practice as embodied knowledge.

In the mid 1990s, the landmark essay collection, *Ground Truth*, noted that any definition of geographic information systems (GIS) or the digital maps produced by GIS should acknowledge the idea that the technology is "part of a contemporary network of knowledge, ideology, and practice that defines, inscribes, and represents environmental and social patterns within a broader economy of signification that calls forth new ways of thinking, acting, and writing" (Pickles 1995: 4). For Haraway, too, maps are clearly ideological and serve social interests: "[M]aps are models of worlds crafted through and for specific practices of intervening and particular ways of life" (1997: 35). Harley (1992: 242) also understood the map as an ideological mode of knowledge production:

> All maps strive to frame their message in the context of an audience. All maps state an argument about the world, and they are propositional in nature. All maps employ the common devices of rhetoric such as invocations of authority. . . . Rhetoric may be concealed but it is always present, for there is no description without performance.

Cosgrove's work (1991, 1999, 2001; Cosgrove and Martins 2000), some of which is of particular interest to this essay, has explored the ways in which cartographic imagery shapes and is shaped by cultural assumptions and influences the geographic imagination. For Cosgrove (1999: 2), mapping is a subjective and communicative practice that constitutes a method for:

> tak[ing] the measure of a world, and more than merely take it, to figure the measure so taken in such a way that it may be communicated between people, places or times. The measure of mapping is not restricted to the mathematical; it may equally be spiritual, political or moral.

Crampton has been concerned with the ideological value of the map as well; he understands the map as bearing agency and invoking cultural contexts. He has expressed concern "with limitations in the ways that populations, locational conflict, and natural resources are represented within current GISs, and the extent to which these limits can be overcome by extending the possibilities of geographic information technologies" (Crampton 2003a: 1). Also relevant to this essay is Schuurman's notion that there exist three "waves" of critique in the history of GIS, of which the publication of *Ground Truth* played a large role. These three waves, Schuurman (2000: 570) notes, reflect both a changing zeitgeist in the GIS landscape and shifting views "on the part of the critic." The first two waves of GIS critique were often fraught with the suggestion that GIS was a positivist, value-neutral technology that functioned as "a mere *tool* of knowledge production" (Propen 2005: 133). Later critiques understood GIS as socially constructed (see Harvey 2001; Harvey and Chrisman 1998), as well as necessarily informed by a feminist theory approach (see Kwan 2002; McLafferty 2002; Schuurman and Pratt 2002).

With these three waves of GIS critique in mind, I wish to call attention to what I take to be a critical juncture in the theory, practice, and popular conception of spatial science technologies. Although we must not forget the earlier positivist critiques, for they are not altogether obsolete, I wish to illuminate instead what appears to be a shift in how we have come to conceptualise visualization practices—a shift that has allowed us to arrive at a point where we might utilize the "remote viewing platforms" once critiqued by Gregory and Haraway for their lack of accountability, for purposes that ostensibly work toward the creation of more sustainable environments by invoking specific local or cultural contexts. This essay advocates for the use of digital mapping tools in such ways that engage and promote environmental sustainability through an approach that understands cartographic practice as embodied knowledge.

Prior to arriving at this critical juncture for spatial science technologies—one that appears to bear much recuperative power for greater awareness around their use—scholars such as Derek Gregory, implicitly aligned with Haraway, rightly suggested that the earlier critiques of mapmaking as a more positivistic endeavour were exacerbated with the advent of the spatial sciences. The advent of the spatial sciences, says Gregory, ultimately gave rise to the infinite and remote viewing platform and its accompanying gaze from nowhere:

> [T]hese high-tech global images that construct the world-as-exhibition in such dazzling display have to be produced from *somewhere*. The subsequent development of GIS has hidden its viewing platforms even more effectively, however, and much of the discussion continues to treat GIS as a detached "science."
>
> (1994: 65)

It is this tendency of the spatial sciences to "construct the world as exhibition" that I wish to reflect upon here. It is important to note that I do not necessarily wish to rehash the older debates, though it is indeed necessary to continue to acknowledge and question this propensity toward the detached "view from nowhere" and subsequently the epistemological consequences of spatial science technologies. What I wish to reflect upon, rather, is the question of how we might keep ourselves in check. For instance, in an age of Google Earth, arguably the quintessential remote viewing platform, and one that may be said to construct the world as exhibition (though we will complicate this notion soon enough), what theories and frameworks might we look toward for an understanding of how to aptly conceptualize the sort of cultural work that such tools ought to accomplish? To this end, I understand the map as both socially constructed and as purporting to represent a "correct" model of the physical world. I contend that such attempts to portray the physical world through cartographic representation ought not to be understood as part of the allegedly positivist project that is cartographic representation; rather, these representations bear the potential to convey partial perspectives, and are well worth interrogating. With these ideas in mind, I advocate a move beyond the earlier positivism debates, in order to rethink mapping as a potentially productive cultural practice—one that, in order to help promote such aspects as environmental sustainability, must invoke and convey a partial perspective.

Cartographic practice as potentially complicit in the totalizing vision

Perhaps the greatest teaching of the earlier GIS debates, at least for our purposes in this chapter, lies not in the primary fact of their allegation, but rather in the implications of their critique. That is, to suggest that GIS has or had positivist underpinnings, which Schuurman (2000: 575), in a thorough discussion of these debates, notes as "encompassing 'assumptions of objectivity, value-neutrality, and the ontological separation of subject and object'," is to suggest that the tools and artifacts produced by GIS lack accountability for the knowledge claims implicit in their work. This potential for lack of accountability also exemplifies what Harding (1996: 17) once described as the intellectual shortcomings inherent in the "conventional conceptual frameworks of the natural and social sciences." In the context of a discussion about science studies, or more specifically, what has come to be known as the science wars, Harding (1996: 17) notes that institutions within the natural and social sciences tend to "produc[e] the kinds of information useful to the administrators and managers of nation states, multinational corporations, and militaries." It is fair to note that Harding was not referring specifically to mapping projects with this statement, and as Schuurman (2000: 570) has noted, "[s]ociological studies of science and technology predate discord about GIS." Nonetheless, Harding's critique may still be viewed as applicable.

Of particular interest to this essay is Harding's (1996: 16) questioning whether the natural and social sciences are capable of "producing the kinds of knowledge that are needed for sustainable human life in sustainable environments under democratic conditions." A little more than a decade later, we may pose this question once again. Modern technoscience, she says, has been of little help in forwarding environmental sustainability; we might attribute this dilemma, she contends, to the fact that "representations of nature, society, and maximally effective knowledge production in the sometimes more effective 'local knowledge systems' conflict with those in modern technosciences" (Harding 1996: 17). The environmental sciences are then faced with the additional challenge of having to negotiate "between the principles of these modern sciences and of both local and social knowledge of environments" (Harding 1996: 18). I propose that by understanding cartographic practice as embodied knowledge, fields such as environmental science and the institutions encompassed by them may use the tools of technoscience to help create more sustainable environments.

Also concerned with the potential lack of accountability inherent in modern technosciences and consistent with Harding's critiques, Haraway's contention with the practices of contemporary visualization (such as those, perhaps, constituted by GIS) focuses on the propensity toward disembodiment, and the consequent lack of responsibility toward the subjects represented through these practices:

> The instruments of visualization in multinationalist, postmodernist culture have compounded these meanings of dis-embodiment. The visualizing technologies are without apparent limit. . . . Vision in this technological feast becomes unregulated gluttony; all perspective gives way to infinitely mobile vision, which no longer seems just mythically about the god-trick of seeing everything from nowhere, but to have put the myth into ordinary practice.
>
> (Haraway 1992: 189)

One example of an iconic, cartographic image produced by spatial science technologies that may, upon first reading, be understood as constructing the "world-as-exhibition" is NASA's (1972) photograph AS17–22727, taken during Apollo's final journey in December of 1972 (Figure 7.1). On the one

Figure 7.1 NASA Photo AS17–148–22727 taken December 7, 1972. Source: http://spaceflight.nasa.gov/gallery/images/apollo/apollo17.

hand, Cosgrove (2001: 260, emphasis added) seems to implicitly understand this photo as an object of cartographic practice when he notes that "the image's *geographical*, compositional, and tonal qualities give it unusually strong imaginative appeal, aesthetic balance, and formal harmony." But is to refer to the image as "geographical" in nature necessarily to consider it an object of cartographic practice? That is, Cosgrove (2001: 261) subsequently notes that:

> the frequency with which photo 22727 is reproduced . . . suggests that its status is iconic rather than cartographic. While it is instantly recognized as an image of the earth, few register its precise geographical contents. Most respond primarily to its cosmographic and elemental qualities.

David Turnbull (1989: 1), too, complicates the definition of the map, asking: "[w]hat are maps and what are their function? What is the difference between a map and a picture? What is the relationship of the map to the landscape it represents? How do you 'read' a map?" Turnbull also notes that maps are generally either *iconic* or *symbolic*. An iconic representation attempts "to directly portray certain visual aspects of the piece of territory in question," whereas a symbolic representation might tap into cultural or social norms or expectations, or make use of "purely conventional signs and symbols, like letters, numbers, or graphic devices" (Turnbull 1989: 3). Many map genres (such as feature maps, topographic maps, news maps, and so on) implicitly employ both iconic and symbolic features. Even further, Cosgrove and Martins (2000: 110), in discussing postmodern, performative mapping, note that "mapping images do not have to be accurate or even realistic, and spatial expressions can be as inventive as possible, showing impossible angles, views, colors, resolutions and situations." Invoking the work of Pickles, they describe how images of the globe in particular have become so heavily laden with symbolic power that "their 'shadows' can be presumed in images which contain no map form at all, relying for meaning on preformed mental stocks of global and topographic imagery to trigger messages and associations" (Cosgrove and Martins 2000: 110). Given the status of photo 22727 as an iconic, geographic image, I suggest that the photo may indeed count as a map, or as an object of cartographic practice and knowledge. If we recall Cosgrove's definition of the map, as mentioned earlier, we may see that iconic, geographic images such as photo 22727 may accomplish the sort of cultural work that influences our understanding of the world and shapes the geographic imagination:

> To map is in one way or another to take the measure of a world, and more than merely take it, to figure the measure so taken in such a way that it may be communicated between people, places or times. The measure of mapping is not restricted to the mathematical; it may equally be spiritual, political or moral.

> (Cosgrove 1999: 1–2)

Thus, in moving away from the more traditional notion of cartography as positing neutral, correct, relational models of the terrain, we may expand the definition of the map to include and promote those geographic, spatial representations that foster or encourage greater empathy and critical consideration for the worlds in which we reside. Cosgrove also understands mapping as a knowledge-making practice that encourages and enables critical thinking and the ability to step outside of our traditionally held assumptions in order to arrive at new imaginings of our world:

> Acts of mapping are creative, sometimes anxious, moments of coming to knowledge of the world, and the map is both the spatial embodiment of knowledge and a stimulus to further cognitive engagements.
>
> (1999: 1)

And as we will soon see, interpretations of photo 22727 may be understood as enabling cognitive engagement with the idea of how we understand or envision our world.

Reading the cultural work of NASA photograph 22727

Upon an initial reading of photo 22727, NASA may very well be understood as having rooted their "condition of being heard to make rational knowledge claims" in a "politics of universality," not partiality (Haraway 1992: 195). On the one hand, photo 22727 may very well be the epitome of the "gaze that mythically inscribes," as it quite literally depicts a view from above; on the other hand, we may also understand the image as invoking a view from the body, or a partial perspective. To view photo 22727 is to see that it dramatically displays Earth as a singular entity, surreal and lacking context. The photo presents the viewer with the "whole, unshadowed globe floating in the blackness of space and given NASA number AS17–22727" (Cosgrove 2001: 257). Viewed as such, the photo may be interpreted as disembodied, as perpetuating an illusion of "infinite vision" (Haraway 1992: 189). Or, perhaps more in accord with the idea of the photo as a mapped space, we might consider its scale, which seemingly delimits a territory encompassing "the whole of creation" (Cosgrove 1999: 2). That is, Cosgrove notes:

> [i]n scale, mapping may trace a line or delimit and limn a territory of any length or size, from the whole of creation to its tiniest fragments; notions of shape and area are themselves in some respects a product of mapping processes.
>
> (1999: 2)

The broad delimiting of this mapped image may initially contribute to an interpretation of its being rooted in a politics of universality, not partiality. Such a reading might then place it roughly along the register of what Cosgrove

(2001: 263) terms the "one-world discourse" interpretation of the image. More specifically, Cosgrove (2001: 262–3) notes that interpretations of photo 22727 have generally been framed by two "related discourses": what he terms the "whole-earth" discourse, and the "one-world" discourse. The one-world discourse, he feels:

> concentrates on the global surface, on circulation, on connectivity, and communication. It is a universalist, progressive, and mobile discourse in which the image of the globe signifies the potential, if not actual, equality of all locations networked across frictionless space. Consistently associated with technological advance, it yields an implicitly imperial spatiality, connecting the ends of the earth to privileged hubs and centers of control.
>
> (Cosgrove 2001: 263)

Upon further consideration, however, the photo may consciously or unconsciously invoke a dissonance in its viewer. The viewer comes to understand its gaze as complex, destabilizing, and even humbling. Perhaps, then, the image is not disembodied, per se, but rather, decentred:

> [T]he image . . . radically destabilizes the cultural part of the conventional meaning of Earth. Not only is the earth separated from the idea of consciousness in the "dead" space beyond its surface, but it is no longer regarded as primarily the "home of Man." Earth is viewed as having an intrinsic life, even its own intelligence as a homeostatic system, and all of its different species accorded dignity equal to that of humans. Humanity is decentered, and by regarding humans as merely one among a multitude of species the cultural variety which is a distinctive feature of our species is suppressed.
>
> (Cosgrove 1991: 128–9)

In this sense, then, for Cosgrove (2001: 261), the "absence of cultural signifiers . . . challenge[s] Western humanism's long-held assumption of superiority in a hierarchy of life." The mapped image and its broad scale may then invoke in the viewer or symbolize a sense of holistic responsibility and kinship, as opposed to distance or disembodiment. Such a reading of the photo might then be more closely aligned with the "whole-earth" discourse, which emphasizes:

> the globe's organic unity and matters of life, dwelling, and rootedness. It emphasizes the fragility and vulnerability of a corporeal earth and responsibility for its care. It can generate an apocalyptic anxiety about the end of life on this planet or warm sentiments of association, community, and attachment.
>
> (Cosgrove 2001: 262–3)

In other words, a "whole-earth"-oriented reading of the photo might understand the mapped image as perpetuating "an objective vision that initiates, rather than closes off, the problem of responsibility for the generativity of all visual practices" (Haraway 1992: 190). Functioning as a mapping, then, the photo may "measure, trace, and represent spatiotemporal concepts and connections" (Cosgrove and Martins 2000: 97). And rather than understanding the photo as disembodied and totalizing, the photo may perhaps be understood as partial and embodied—it may function symbolically to invoke the viewer's schema or social contexts relative to the cultural moment in which they are immersed.

Perhaps, then, the paradox of photo 22727 is its ability to simultaneously distance and engage the viewer, a paradox conveyed no more ironically than by the fact that this seemingly decontextualized image of Earth floating in space has become an icon for the global environmental cause. It is also interesting to note that by decentring humanity and "regarding humans as merely one among a multitude of species" (Cosgrove 1991: 128–9), the photograph works against the anthropomorphic vision. Likewise, the photo destabilizes the "long-evolving Western global image," and as a result, "the world is radically decentered" (Cosgrove 2001: 261). By decentring Europe and affording privilege to the South, the photo works against the historically ethnocentric view of the globe that Monmonier (1996) and others have also often critiqued.

The cultural work accomplished by photo 22727 not only demonstrates Cosgrove's (1991: 130) view that "geography's words and images have always had a certain power to construct as much as to reflect the orders which it represents," but is also a powerful reminder of the ways in which spatial science technologies may function to construct multiple orders, vantage points, or visions for their subjects. And as we construct these images, we must consider the implications of their perspective.

As mentioned earlier, Haraway's insistence on the "embodied nature of all vision" calls for a reclaiming of the "sensory system that has been used to signify a leap out of the marked body and into a conquering gaze from nowhere" (1992: 188). In other words, the spatial science technologies responsible for producing these images of our world ought not to distance the viewer from the image; instead, we might utilize these systems in such a way that more holistically engage the mind and body of the viewer, making the viewer more fully accountable for what they see. It is my contention that a reclamation of this sensory system necessarily entails a critique of the traditionally hegemonic ideologies of abstraction and detachment potentially carried out by technologies such as GIS, and thus a subsequent re-visioning of the cultural practices that inform spatial science technologies in order to make such imagery possible.

It is also important to note however, that our contention ought not to be solely with the technological apparatus that makes the image possible. And of course, for Haraway, it is not. Like Latour and Woolgar (1986), Haraway understands the instrument and its artifact—or, let us say the GIS and its map, for example—as inextricably linked to the social contexts in which they are situated. Nonetheless, the instrument, the artifact, and its context are equally implicated in meaning-making: "Technologies and scientific discourses can be partially understood as formalizations, i.e., as frozen moments, of the fluid social interactions constituting them, but they should also be viewed as instruments for enforcing meanings" (Haraway 1992: 164). We might then think of technology, scientific discourse, and their artifacts as culturally constructed—not news to those familiar with the later GIS critiques, which considered GIS from the vantage point of social construction and feminist theory.

Again, though, how might we ensure that we act responsibly to utilize these culturally constructed discourses, instruments, and their artifacts to create sustainable worlds? How do we avoid the seeming propensity toward disembodied, totalizing visualization practices, especially when GIS technologies have the potential to appropriate the "natural" or physical world for the purposes of "infinitely mobile vision" without self-regulation or critical thought? To this end, Haraway (1992: 188) again advocates for the "embodied nature of all vision." Embodiment in this sense need not necessarily be "organic," and may incorporate "technological mediation" (Haraway, 1992: 189). In other words, embodiment does not necessarily imply corporeality, and may involve technology—the two are not mutually exclusive. With these ideas in mind, we might look to digital mapping technologies as providing a point of entry for understanding cartographic practice as embodied knowledge.

That is, if we understand the GIS map as a technologically-mediated representation of the natural world, we might view mapping as a cultural practice that bears the potential for embodied knowledge. The map will likely always be a cultural construction that, to some extent, attempts to portray a "correct" model of the physical environment; however, such attempts at more realist representations of the earth need not necessarily bear a negative connotation. Not, I would argue, if such attempts, seemingly a necessary component of modern technological sciences, also incorporate accountability, responsibility, critical thought, and embodied knowledge—if they incorporate and implement a partial perspective. If we may represent by way of partial perspective rather than totalizing vision, then the game changes.

Cartographic practice as initiating partial perspective

As we have seen, a closer reading of photo 22727 necessarily complicates the image, allowing for its interpretation as both totalizing and partial. On the one hand, the photo arguably enables the "modern decorporealization of vision," in that the photo's remoteness and lack of accountability make it almost immaterial, intangible, and even surreal. In this sense, the image may be read to invoke what Haraway (1992: 188) refers to as "the gaze that mythically inscribes all the marked bodies, that makes the unmarked claim the power to see and not be seen, to represent while escaping representation." On the other hand however, the fact that such a surreal or allegedly disembodied image has become "a favored icon for environmental and human-rights campaigners" (Cosgrove 2001: 261) indicates that, on some level, it creates meaning that enables "a chance for a future," and as Haraway (1992: 187) notes, "[w]e need the power of modern critical theories of how meanings and bodies get made, not in order to deny meaning and bodies, but in order to live in meanings and bodies that have a chance for a future." If photo 22727 does indeed serve to construct a meaning that enables a chance for a future—and we have shown that it may be interpreted in such a way—then the question of how it does so becomes a complex one. That is, the photo's understated power may be read as engaging the viewer's individual, global environmental schema through its ability to invoke a holistic vision of a "global" environmental cause. Again, the interesting point here seems to be that within the mythical gaze of photo 22727, there also resides a partial perspective.

For Haraway (1992: 191), the notion of partial perspective offers more than the relativist position for which we might initially mistake it. As she puts it, "[r]elativism and totalization are both 'god-tricks' promising vision from everywhere and nowhere equally and fully . . . [b]ut it is precisely in the politics and epistemology of partial perspective that the possibility of sustained, rational, objective inquiry rests." Harding, whose understanding of objectivity may be understood as informing Haraway's notion of partial perspective, aligns herself with the field of science studies when offering a similar caution against viewing relativism as an antidote for the sort of objectivity that we might take to be instrumentally-minded: "Science studies does not claim that sciences are epistemologically relative to each and every culture's beliefs such that all are equally defensible as true. Rather, the point is that they are historically relative to different cultures' projects" (Harding 1996: 17). Likewise, for Haraway, the notion of partial perspective does not entail endless relativism; rather, it requires the ability to translate what we see, specific to the cultural moment at hand, and consider and apply that vision in terms that pertain to particular projects other than our own:

We need to learn in our bodies . . . how to attach the objective to our theoretical and political scanners in order to name where we are and are not, in dimensions of mental and physical space we hardly know how to name. So, not so perversely, objectivity turns out to be about particular and specific embodiment, and definitely not about the false vision promising transcendence of all limits and responsibility. The moral is simple: only partial perspective promises objective vision. This is an objective vision that initiates, rather than closes off, the problem of responsibility for the generativity of all visual practices.

(Haraway 1992: 190)

In other words, the idea of partial perspective may be understood as requiring socioculturally specific compassion and empathy. It is, perhaps, about achieving a sort of objectivity that is rooted not in the claims of an allegedly neutral or utilitarian science, but of understanding and advocacy.

And as Cosgrove notes, the astronauts who first witnessed the view that later became known as photo 22727 did seem to experience this more reflexive sort of objectivity, one more akin to embodied knowledge, even though they likely did not refer to their experience explicitly in such terms. As Cosgrove describes the photo's situatedness by removing the veil of photo 22727's gaze and also mentioning "those few humans" who were on board Apollo 17 when the photo was taken, we may once again understand the photo as conveying not so much a mythical gaze but more so a partial perspective, albeit a privileged perspective, nonetheless:

> Those few humans who actually witnessed the revolving terracqueous globe and who produced photo 22727 describe their experience in terms of awe, mystery and humility. The axis of world order, if it existed for them, stretched infinitely above and below the global surface.
>
> (Cosgrove 1991: 130)

In addition to understanding the photo in terms of the "whole-earth" discourse around which it may be interpreted, Cosgrove's mention of the photo's "producers" who "describe their experience" works to ascribe further accountability to what might otherwise be read as a disembodied or infinite gaze. That is, even though accountability may only be ascribed to "those few humans," it may be delegated nonetheless. And it is precisely this sort of accountability that I believe makes all the difference in discussions of how spatial science technologies work to construct and communicate their perspective. Understanding the photo not only through the lens of the whole-earth discourse, but also from the vantage point of its producers' experience, allows us to understand the photo as producing a very specific sort of local, embodied knowledge. Thus, to understand spatial science technologies and the artifacts they produce in terms of partial perspective means understanding visualization practices in terms of their potential for producing specific ways of seeing:

> The "eyes" made available in modern technological sciences shatter any idea of passive vision; these prosthetic devices show us that all eyes, including our own organic ones, are active perceptual systems, building in translations and specific *ways* of seeing, that is, ways of life. There is no unmediated photograph or passive camera obscura in scientific accounts of bodies and machines; there are only highly specific visual possibilities, each with a wonderfully detailed, active, partial way of organizing worlds.
>
> (Haraway 1992: 190)

We may now understand photo 22727 as implicitly communicating situated, embodied knowledge, even though this particular image does quite literally depict a view from above. Is it enough, however, to understand an image as *implicitly* conveying a "view from a body" (Haraway 1992: 195) or should we expect more from the cultural work accomplished by these images and the technologies that produce them? Because there is likely no easy answer to a question such as this, I propose instead a sort of compromise; the idea that interpretations of what counts as partial perspective may take place along a sort of continuum—a continuum that accounts for, along its register, versions of more or less explicitly communicated embodied knowledge. To take this idea further, now that we understand an image such as photo 22727 as bearing the potential for embodied knowledge, we might also push the envelope and ask whether the image conveys a partial perspective to the full extent that we might hope. In other words, our compromise ought not to lower our expectations for a more explicit or holistic expression of embodied knowledge. Haraway (1992: 169), too, pushes for greater accountability on the part of all groups involved with the use of these technologies, when she asks:

> What kind of constitutive role in the production of knowledge, imagination, and practice can new groups doing science have? How can these groups be allied with progressive social and political movements?

With this call to action in mind, it becomes clearer that the groups developing and utilizing the tools of spatial science technologies must consider the potential for alliances with the sort of progressive social, political, and environmental movements to which Haraway refers. And conversely, I might add that "progressive social, political, and environmental movements" must also think critically about their intended use of spatial science technologies. Recent uses of Google Earth by environmental organizations, for example, provide an interesting starting point.

Higher expectations for visualization practices: Toward a more explicit partial perspective

Google Earth is a virtual mapping tool that "drapes satellite imagery over 3-D topographic data" (Dicum 2006: 1). Users can create layered maps that show specific relationships between places, events, or artifacts, and then "share their explorations with others" (Google Earth 2007: 1). Because the application creates the illusion of flying through a landscape, it engages the body of the user in the way that other mapped representations might not, essentially allowing users to "pilot" their own experience, in turn giving them "a very intimate understanding for how a place is laid out" (Dicum 2006: 1). Because the tool allows users to experience a very specific perspective, it arguably advocates a "particular and specific embodiment," as opposed to the promise of "transcendence of all limits and responsibility" (Haraway 1992: 190). Even further, because Google Earth is designed to create such specific, embodied perspectives while also showing relationships between places and other markers, the software application may be used as a persuasive object of visual rhetoric—one that may function in the production of specific types of embodied knowledge.

Working in the positive, I suggest that Google Earth has the capacity to enable the creation or expression of the sort of partial perspective for which Haraway advocates—one that moves beyond the type of knowledge work accomplished by images such as photo 22727. It is important to note however, that although such capability is due in part to the technology's functionality, it is due largely to the interest of Google Earth and its developers, considered perhaps to be these "new groups doing science," in allying themselves with "progressive social and political movements" (Haraway 1992: 169). The success of tools like Google Earth in creating more thoughtful cartographic representation is also a function of the overtly critical intent conveyed by the environmental groups who have recently appropriated the tool in an effort to create their own embodied, cartographic representations.

That is, much to the interest of this essay, Google Earth has recently been taken up by environmental groups in such a way that allows for what seems to be a new sort of visual, environmental rhetoric. This appropriation of Google Earth by environmentalists is one example of how we might use spatial science technologies and spatial data to forward partial perspectives and thus better understand cartographic practice as embodied knowledge.

A recent article published in the online edition of the *San Francisco Chronicle* describes the appropriation of Google Earth by environmental groups such as the Sierra Club, suggesting that the software application "could change the power balance between grassroots environmentalists and their adversaries" (Dicum 2006: 1). The Sierra Club recently used Google Earth to help forward their agenda to protect the U.S. Arctic National Wildlife Refuge. They created what is referred to as a Google Earth "annotation," which allows users to virtually visit the Refuge. In doing so, says the Sierra Club, users can:

> get out there and see why this place is worth protecting . . . . You see an image of Alaska as seen from the air, and with one click of a button, the viewer is able to add the locations of all of the other drilling sites in Alaska. It really drives home that most of Alaska is already open to oil and gas development and there's this one place that we've managed to protect thus far. This kind of visual perspective on environmental problems transforms vague policy debates into concrete problems.
>
> (Dicum 2006: 1)

Here we see several issues at play. First, the purpose of the tool is to engage the body in such a way that affords a particular vision—a vast vision, perhaps, but not necessarily limitless or totalizing. Next, we see the user immersed in a technologically mediated representation of the physical environment. In this sense, Google Earth functions as a sort of technological artifact that is not only socially constructed in the relationships it conveys, but one that also purports to represent a "correct" model of the physical world—correct, that is, relative specifically to the context in which it is immersed. Finally, this cartographic representation may be understood as partial and interested—it makes use of cartographic representation in such a way as to produce embodied knowledge that, in this case, forwards an environmental agenda.

Conclusion

Applications such as Google Earth are not by any means impervious to critique. In a somewhat related discussion of the tools used in community mapping projects, for example, Perkins (2007: 134–5) notes that:

> [t]echnological advances in the last five years have led to new community mapping initiatives that aim to build collaborative, community-led alternatives to commodified map data. Many of these initiatives have exploited high resolution satellite data and mapping from portals such as Google Maps or Google Earth. These hacks and mashups, however, still depend upon the commercial provision of base map and image data.

Although tools such as Google Earth are indeed products of corporate technoscience, and may be understood as perpetuating the corporate commodification of information, they are, undeniably, explicitly enabling the expression of embodied, cartographic knowledge.

Rebecca Moore, a software developer who works on Google Earth, understands it as an advocacy tool that can positively impact the way we understand our world. Of the application, she says:

> I think that this has the potential not only to raise people's environmental consciousness but to raise their consciousness of humanity . . . I see it as making the world a smaller place in a good way; giving everyone a greater intimacy with the Earth and the rest of the people and the plants and animals that share it with us.
>
> (quoted in Dicum 2006: 4)

In raising not only our environmental consciousness but also our consciousness of humanity, perhaps this tool and others like it may function as purveyors of embodied cartographic knowledge—as mapped representations that communicate a partial perspective.

In this essay I have sought neither to praise nor disparage one particular mode of technological visualization over another, nor have I sought to identify the silver bullet that will resolve the dissonance we experience as we discover and make use of new applications for spatial science technologies. Rather, I have first sought to consider the implications of a once allegedly value-neutral technology, and then complicate this notion of the totalizing view, arguably an outcome of such neutrality, by searching for accountability where seemingly there was none—where quite literally, with mapped images such as photo 22727, there was a view from above. At this point, then, it becomes necessary to push the notion of accountability even further; holding it to a higher standard of overtly communicated critical intent in the purveyance of embodied cartographic knowledge. In other words, perhaps I have sought to identify a continuum of accountability in the production of cartographic knowledge—a continuum that has on its register the relative ability of cartographic representation to engage not only the mind, but also the body, in its ability to foster new schemas within the geographical imagination.

Although recent scholarship in critical GIS, for example, has considered the implications of spatial science technologies for such aspects as our right to privacy, both in terms of the dissemination of personal data and the potential for constant locatability (Crampton 2003b; Monmonier 2002; Propen 2005), and although I believe it is necessary to continue to examine such issues, I have had a different sort of question in mind with the writing of this essay: how might we use cartographic knowledge to learn to live with and understand one another with compassion and with empathy? In short, how might we use cartographic knowledge to create more sustainable and liveable worlds? Rebecca Moore's understanding of Google Earth is able to account for these questions to an extent. By understanding and using the tool in such a way as to "raise people's environmental consciousness . . . [as well as] their consciousness of humanity," Moore implicitly understands Google Earth as enabling the expression of embodied cartographic knowledge (Dicum 2006: 4). Whether the Sierra Club's appropriation of Google Earth counts as local mapping, participatory GIS, or even a hybrid of the two, although an interesting question, becomes not so much the point. However we might define or constitute such organizational pairings (perhaps a legitimate question left open with this essay), the Sierra Club's use of Google Earth nonetheless takes a step toward answering Haraway's call to action. Remember, Haraway (1992: 169) asks: "[w]hat kind of constitutive role in the production of knowledge, imagination, and practice can new groups doing science have? How can these groups be allied with progressive social and political movements?" I suggest here that the pairing of technologies such as Google Earth with the work of environmental groups such as the Sierra Club demonstrates clearly the ways in which groups developing or working with spatial science technologies may implicitly or explicitly ally themselves with progressive social, political, and environmental movements. Equally important, as we have seen, is the *mode* of knowledge production. By engaging both the mind and body of the user, Google Earth helps enable the creation of a new, embodied, geographical imagination—one that works toward the visualization and understanding of multiple lived worlds.

Embodied cartographic knowledge provides a more intimate understanding of a place, and in doing so, perhaps a closer relationship and feeling of responsibility and accountability toward it. And this, for Haraway, is quite the point. That is, our "pictures of the world should not be allegories of infinite mobility and interchangeability, but of elaborate specificity and difference and the loving care people might take to learn how to see faithfully from another's point of view, even when the other is our own machine" (Haraway 1992: 190). In this case, the other's point of view is most certainly a product of our own machine, and has been allied with progressive social, political, and environmental groups to show us how we might create more liveable, sustainable worlds.

Note

1 Acknowledgments: I am very grateful to Elizabeth Britt, Michael Salvo, and Elizabeth Shea for their feedback on early versions of this chapter, and to the reviewers for their feedback on later versions of the chapter. I would also like to express my great appreciation for the work of the late Denis Cosgrove, whose notion of postmodern mapping informs my own perspective on critical cartographies.

References

Cosgrove, D. (1991) 'New world orders', in C. Philo (ed.) *New Words, New Worlds: Reconceptualising Social and Cultural Geography*, Edinburgh: IBG Social and Cultural Geography Study Group.

Cosgrove, D. (1999) 'Introduction: mapping meaning', in D. Cosgrove (ed.) *Mappings*, London: Reaktion Books.

Cosgrove, D. and Martins, L.L. (2000) 'Millennial geographics', *Annals of the Association of American Geographers*, 90: 97–103.

Cosgrove, D. (2001) *Apollo's Eye: A Cartographic Genealogy of the Earth in the Western Imagination*, Baltimore, MD: Johns Hopkins University Press.

Crampton, J.W. (2003a) *How Can Critical GIS Be Defined?* <http://geoplace.com/gw/2003/0304/0304cgis.asp>.

Crampton, J.W. (2003b) *The Political Mapping of Cyberspace*, Chicago, IL: University of Chicago Press.

Dicum, G. (2006) 'Green eyes in the sky: Desktop satellite tools are changing the way environmentalists work', *San Francisco Chronicle*, January 11, <www.sfgate.com/cgi- bin/article.cgi?f=/g/a/2006/01/11/gree.DTL>.

Google Earth (2007) *Explore, Search and Discover*, <http://earth.google.com/>.

Gregory, D. (1994) *Geographical Imaginations*, Cambridge: Blackwell Publishers.

Haraway, D. (1992) *Simians, Cyborgs and Women: The Reinvention of Nature*, New York: Routledge.

Haraway, D. (1997) *Modest_Witness@Second_Millennium: Femaleman_Meets_ Oncomouse*, New York: Routledge.

Harding, S. (1996) 'Science is "good to think with"', in A. Ross (ed.) *Science Wars*, Durham, NC: Duke University Press.

Harley, J.B. (1992) 'Deconstructing the map', in T.J. Barnes and J.S. Duncan (eds) *Writing Worlds: Discourse, Text and Metaphor in the Representation of Landscape*, New York: Routledge.

Harvey, F. and Chrisman, N. (1998) 'Boundary objects and the social construction of GIS technology', *Environment and Planning A*, 30: 1683–1694.

Harvey, F. (2001) 'Constructing GIS: Actor networks of collaboration', *URISA Journal*, 13: 29–38.

Kwan, M. (2002) 'Is GIS for women? Reflections on the critical discourse in the 1990s', *Gender, Place and Culture*, 9: 271–279.

Latour, B. and Woolgar, S. (1986) *Laboratory Life: The Social Construction of Scientific Facts*, Princeton, NJ: Princeton University Press.

McLafferty, S.L. (2002) 'Mapping women's worlds: Knowledge, power and the bounds of GIS', *Gender, Place and Culture*, 9: 263–269.

Monmonier, M. (1996) *How To Lie with Maps*, Chicago, IL: University of Chicago Press.

Monmonier, M. (2002) *Spying with Maps: Surveillance Technologies and the Future of Privacy*, Chicago, IL: University of Chicago Press.

NASA (1972) *Photo 22727*, <http://spaceflight.nasa.gov/gallery/images/apollo/apollo17>.

Perkins, C. (2007) 'Community mapping', *The Cartographic Journal*, 44: 127–137.

Pickles, J. (1995) 'Representations in an electronic age: Geography, GIS, and democracy', in J. Pickles (ed.) *Ground Truth: The Social Implications of Geographic Information Systems*, New York: Guilford Press.

Propen, A. (2005) 'Critical GPS: Toward a new politics of location', *ACME: An International E-Journal of Critical Geographies. Special Issue: Critical Cartographies*, 4(1): 131–144.

Schuurman, N. (2000) 'Trouble in the heartland: GIS and its critics in the 1990s', *Progress in Human Geography*, 24: 569–590.

Schuurman, N. and Pratt, G. (2002) 'Care of the subject: Feminism and critiques of GIS', *Gender, Place and Culture*, 9: 291–299.

Turnbull, D. (1989) *Maps Are Territories, Science Is an Atlas: A Portfolio of Exhibits*, Chicago, IL: University of Chicago Press.

8 *The 39 Steps* and the mental map of classical cinema

Tom Conley

The field of cultural studies is riddled with the idea of "mapping." The term is often understood to be a superimposition of one or more critical templates of a given discipline onto others from another that can reveal unforeseen patterns or webbings of relations. The student in the field is not so much a cartographer in the historical sense of a surveyor and draftsman than as an intellectual engineer, an individual designing hypotheses affiliated with codes of spatial reason. The analysis of cinema, increasingly associated with cultural studies, also owes much to mapping. Works of given directors are often set over each other in order to reveal pertinent traits and variations that shape their signatures. Film histories are shown as "maps" by which given historical phases are placed adjacent to one other or else overlapped at transitional edges in order to make visible patterns of change or, in a pedagogical sense, to construct grids in which thousands of films can be stored in a vault of memory.

Attention has been drawn less frequently to the idea that films in themselves are maps and that they can be understood as modes of *locational imaging*. It can be observed that in fact many of the films we see insert maps into their visual field and that, as a result, a greater sense of cartography is gained in the relation that the object holds with the image in which it is found. Maps, often vital to the construction of narrative or effects of verisimilitude and historical veracity, can be taken more generally to have much to do with cinematic form. Such is the bolder line of the argument guiding the paragraphs that follow: that cinema is a mapped medium that can be appreciated in a literal (and hopefully, productive) sense through its own cartographic elements. In the same breath, in the wake of Michel Foucault's (1975) and J. Brian Harley's (2001) studies of the ideology of cartography, mapping can be taken to mean how subjectivity is managed through spatial means. From its inception cinema would play a similar role by virtue of obtaining and accruing power in shaping consciousness and identity.

However, students of the medium are quick to show that great films and their creators use the cartographic latency of cinema to raise consciousness about the very power of the medium itself and to "theorize" the process of "mapping" inherent to montage and editing of lexical and visual material.

These directors show their spectators how they are putting them in their "place" for the duration of a projection or emission; how they cause them to determine where they are, not only in respect to the narratives in which they are engaged, but also in a psychogeographical sense, in the consciousness of cognition. Since the advent of narrative in cinema—which is to say, from its very beginnings—maps are inserted in the field of the image to indicate where action "takes place" while, at the same time, they become mental prods that tell (or whisper to) the spectator that they generate fantasies of movement through different places. In a film a map is given to say, "you are here." Yet, in response to the implied declaration, in the active or dialogical relation we hold with the image, we are compelled to respond, "I am elsewhere, I am not where you say I am." From this angle cinema becomes a vital tool in prodding us to "locate" ourselves: to plot a sense of time and place in which the fantasies of a film are set in that dialogue which we bring to its images. Everywhere, it can be surmised, the mental operations of cinema can elicit active contemplation about subjectivity in its determinations of space and place in both individual and collective registers.

Along a narrower line of inquiry I would like to explore these issues from two complementary angles. One, which belongs to philosophies of space and the moving image, opens from the cartographic impulse that inspires Gilles Deleuze's writings on cinema. His is a multifaceted mapping that includes a history and taxonomy of the seventh art, and it is also a philosophy that meshes the process of *thinking* about the nature of being in space with analysis of the moving image. The other, drawing on the same author's way of "reading" the cinema of Alfred Hitchcock, aims at study of the mapped forms of *The 39 Steps*, a signature film in the director's oeuvre, a feature whose cartographies (understood in both geographical and psychic senses) exceeded by far the functional design that would be ascribed to the projections and places that figure in the feature. Analysis of maps in the film will lead to the hypothesis that the locational imaging bears on the theory engaged in the two paragraphs above: that cinema plots spectatorship; that spectators are in a constant process of thinking about where they are, "deterritorializing" themselves, in respect to the power of mapping invested in the medium in general.

Deleuze's most visible theories of mapping are found, first, in his study of space and territory in *Mille plateaux: capitalisme et schizophrénie* (1980, co-written with Félix Guattari) and, second, with terse concision, in a chapter of an essay (originally an article in *Critique* in homage to the late Michel Foucault) in *Foucault* (1986), titled "A New Cartographer." Contemporaneous with Deleuze's writings on cinema, these two essays constitute a theory and a practice. In accord with the general tradition of the *ars poetica*, practice precedes theory. In *Mille plateaux* two types of space, each enveloped in the other, allow us to apprehend and to chart the world's extension. What he calls an *espace lisse* or "smooth space" is without line or border. It bears resemblance to oceans or deserts, is of no easily calculable limit, and is hypothetical in nature in the world we inhabit. It is a space in which thought catalyzed by its very idea leads to wandering and to nomadism. No matter what cardinal points of reference may be used to locate a smooth space, it nonetheless is sensed only through the passage or journey drawn through and about it:

> In smooth space the line is a vector, a direction, and not a dimension or a metric determination. It is a space constructed by local operations with changes of direction. These changes of direction can be owed to the very nature of the pathway taken (. . .); but also to the variability of the goal or the point to be reached.
>
> (Deleuze and Guattari 1980: 597)

Smooth space is "directional, not dimensional or metric," and it is taken up more with events or *haeccities* (sensations of surface) than formed and perceived things (Deleuze and Guattari 1980: 598). It is charged with affect and is sensed haptically, with touch and tactility as much as through sight. Deleuze calls it *intensive*—wrought with intensities—and not extensive, that is, marked with measured distances. "That is why smooth space is occupied by sounds, noises, winds, forces and tactile or sonorous qualities such as in the desert, the steppes, or the glacial regions" (Deleuze and Guattari 1980: 598).

The sea, which would be smooth, is nonetheless striated by astronomical and geographical points of reference, and thus from it emerges "the map, which crisscrosses meridians, parallels, longitudes and latitudes" (Deleuze and Guattari 1980: 598). The sea is the first of all smooth spaces to be gridded with navigational lines, with rhumbs, and to be dotted with compass roses that gather lines at their hub and cause them to radiate. The philosopher adds that *to think is to travel* in the spaces that voyage engenders, and that the motion of voyage simultaneously "deterritorializes" and "reterritorializes" the traveler (Deleuze and Guattari 1980: 602). To see a film, he implies, can entail taking a haptic voyage over its multiple surfaces. It can engage, in an enthusing intransitive sense, unforeseen ways of "becoming" or experiencing. Such is what a consummate director does in and across the films of his or her signature. If indeed most films are about voyages or discoveries, the model of striated and smooth spaces applies immediately to the screened image. Since striation is basically optical in nature, a purely haptic cinema would be of an impoverished visuality, blind, or else so textured that reference to a real world would be fleeting at best. In cinema an extreme close-up would turn an optical form (as it will be shown, one hand grazing another) into a tactile abstraction:

> Where vision is near, space is no longer visual, or rather the eye itself [has] a haptic and not an optical function: no line separates sky from earth, that are of the same substance; there is neither horizon nor

background, perspective nor limit; neither contour, form, nor center; there is no intermediary distanced, or else all distance is intermediary.

(Deleuze and Guattari 1980: 616)

Thus striated space cannot really be called local or topographic, nor can its smooth counterpart be assumed to be global (or cosmographic) because the close-up can turn something local into something absolute.

In these pages it is implied that the screen can be seen in variously "smooth" and "striated" conditions, and that as a result "reading" and "seeing" a film in the same act of cognition resembles the haptic and optical voyages that maps offers to their viewers. In his essay on Foucault Deleuze argues that mapping can be understood in terms of a distinction between an *archive* and a *diagram*. The archive is implicitly likened to an atlas, to an accumulation of "visible" and "discursive" formations that have dictated how the ambient world is seen and deciphered. An archive is an assemblage of items of a genre (a collection of maps or movies) constituting the rudiments of a history with and against which new forms—diagrams—are fashioned. Unlike an archive, a diagram is plotted to shape behavior and to open inherited spaces onto a plane of becoming, in other words, to inspire uncertain travel in spaces that can be engaged crosswise (against the grain of striations) or so as to be smooth enough to invent new and unforeseen relations. "The history of forms [where forms equal 'striations'], an archive, is doubled by a becoming of forces, a diagram" (Deleuze 1986: 50). No matter how it may be scored in appearance, the diagram offers a possibility of plural itineraries to be taken through and beyond the "archive" of our world. "[N]o diagram fails to bear, adjacent to the points it connects, other points, relatively free or unbound, points of creativity, of mutation, of resistance" (Deleuze 1986: 51). In the progression from creativity to resistance a greater degree of freedom gives way to a lesser degree, and a condition of smoothness to one of invention and passage within closed or repressive conditions.[1] A politics of mental mapping emerges. The couple and coupling of archive and diagram implies a good deal about the cartographic impulse of cinema—if it is agreed that films, when they are discerned in their "tradition" (as genres or of given "qualities") are archives, and that those which open other spaces through their play of sound and image are diagrams. Experimental and independent films would be diagrammatic but so also would classical films when displaced into other traditions or times and places. "Deterritorialized," older films would foster creative innovation because they are plotted in what appear to be, in their past time, the uncommonly smooth spaces of their own invention.

These two sites where Deleuze reflects on cartography also figure, albeit in a different lexicon, in *Cinéma 1: L'Image-mouvement* (1983) and *Cinéma 2: L'Image-temps* (1985). In the foreword to the first volume he claims that his is not a history but, rather, a "taxonomy, an essay on the classification of images and signs" of cinema (Deleuze 1983: 7), based on Charles Saunders Peirce and Henri Bergson, with the aid of whom he draws a line of divide between the regime of the "movement-image" (that both moves and moves the sensory apparatus of the viewer) and the "time-image" (tending to be more autonomous in nature) that breaks away, and "exceeds the relations" (p. 283) established between the spectator and the film by locating the image in duration. Movement-images and time-images are interrelated and even enveloped in one another much in the same way as smooth and striated spaces or archives and diagrams. Yet, despite the broad lines of Deleuze's taxonomic categories, the disposition of the two-volume study suggests that World War II and its aftermath bear so much impact on the state of the image that history determines how and why the movement-image gives way to the time-image.

Deleuze's line of demarcation is drawn parallel to what Jean-Luc Godard has noted of the state of the cinematic image in 1945. The Swiss director, observing that director Georges Stevens had concealed the traumatizing 16 mm color footage he had shot of Auschwitz and Ravensbrück while chronicling the end of World War II (Stevens served in the U.S. Signal Corps), was revulsed by what he had witnessed. Confining the film to the basement of his home, he internalized what, more broadly, Godard has taken to be a crisis of the image. What the film documented was in such deficit in respect to the horrors it recorded that the filmmaker was led to wonder if, as Paul Celan said of poetry, images could be fashioned after the fact of the camps. In his *Histoire(s) du Cinéma* Godard (1998: 131–4) boldly reiterates the point in superimposing a shot of a gaunt face from Stevens' chronicle onto an image of Montgomery Clift, supine, in the lap of Elizabeth Taylor, in Stevens' *A Place in the Sun* (1951), prior to inserting photograms of the child-protagonist of Rossellini's *Germany, Year Zero* (1947) who looks at what would seem to be incomprehensible horror. All of a sudden Giotto's portrait of Mary Magdalene (turned at a right angle so as to look down upon the image) appears as if she were looking at a second image from the 1951 feature that Stevens adapted from Dreiser's *An American Tragedy*. Implied is that Giotto's art, when set in the medium of cinema, resurrects what would have been the end of the image, the very image that Hollywood repressed in its sanitized cinema of the post-war years.[2] Godard's question about the camps having changed the state of the image finds its correlative in Deleuze's "archive" that chronicles the end of the movement-image.

The meridian that Deleuze draws along the axis of 1945 separates two different models of film. Yet at the beginning of the project he writes that most cinema we see is pointless despite the "incomparable economic and industrial consequences" that force "the *great* auteurs of cinema" (Deleuze 1983: 8, emphasis added) to be a fragile and threatened species. He states too that his words are not a legend to illustrations in his text; he would prefer his words merely to be "an illustration of the *great* films whose emotion or perception everyone of us more less remembers" (Deleuze 1983: 8, emphasis added). Great cinema risks desecration and extermination; its own history, like that of the camps, attests to fragility and destruction in view of the events of the twentieth century. He takes his words to be illustrations, verbal images seen and read at once—like a map—to inspire affective motion and, at the same time, as does Godard, to turn an archive of cinema into a diagram. *Great* films and *great* auteurs "think," he adds, with movement-images and time-images instead of with concepts (Deleuze 1983: 7–8).

From 1945 onward the "crisis" of the movement- or action-image owes to five causes. Unanimity (1) is lost, and so also the bird's eye view of the city; no longer "seen from on high, the city standing and erect with skyscrapers in counter-tilt," becomes the "flattened city, the horizontal city at the height of any person" (Deleuze 1983). The end of the global or synthetic situation gives way (2) to haphazard arrangements of events that proliferate, no longer in a narrative webbing, and that merely happen to take place. The cinema (3) begins to wander on its own, passively inscribing action at the centre and peripheries of the frame, in "any-spaces-whatsoever" (*espaces quelconques*), which include warehouses, urban detritus, and the "tattered fabric of the city" and not the integral spaces of earlier realism. (4) Clichés abound in the shape of "floating images . . . that circulate in the outer world but that also penetrate each and every one and constitute their inner worlds" (Deleuze 1983: 281) to the degree that characters (and spectators) can only think and feel with clichés. Finally (5), the world becomes an immense conspiracy, no longer engineered by a single magical force or agent, such as René Clément's Professor Crase in *Paris qui dort* (1924) or the mad genius of Lang's *Testament of Dr. Mabuse* (1933). Surveillance, something of an avatar of GIS systems and satellite locational apparatus, is found everywhere. What Deleuze finds so remarkable in these five aspects in the films of Robert Altman and Sidney Lumet has origins in Italian Neo-Realism, especially *Roma, città aperta* (1945) and *Païsa* (1947), *The Bicycle Thief* (1949), and (with greater deference to the time-image), *Voyage to Italy* (1951).

The impact is felt later in France, he argues, because the ambiguous conditions of the Occupation "did not favor a renewal of the cinematographic images that had been stuck in the frame of a traditional action-image and at the service of a properly French 'dream'" (Deleuze 1983: 285). The *Nouvelle vague* acceded tardily to new cinema, and more by way of its discovery of Howard Hawks and, notably, of Alfred Hitchcock for whom "mental images" were of the order of Peirce's concept of "thirdness," that is, a "term referring to a second term by the intermediary of another or other terms" (Deleuze 1983: 266) in a sort of *relation* in which a situation is thought through by virtue of the spectator looking at characters who gaze and reflect upon the situations in which they are found.[3] The mental image is conveyed through the sight of ordinary objects that become sites where many relations converge. They are signs of a tangle of "symbolic acts" and "intellectual feelings" (Deleuze 1983: 268). For Deleuze the French adepts of Hitchcock have studied the director's style not to plot relations but to use them to form a new cinematic substance. With Hitchcock they succeed in making film *think*, and even having it engage formidable creative challenges in the wake of the crisis of the movement-image. It is through the British director's mental image that a new "thinking image" (Deleuze 1983: 290) is conceived.

Surface tensions—where letters, names, objects, and passages move about the frame—evince the relational or mental quality of the image for the observer who looks at Hitchcock's films.[4] The director is the master traumatizer both in his British (pre-war) and American (post-war) phases by either anticipating or explicating the greater crisis of the image when it befalls cinema in 1945. Hindsight tells us that the conspiracies that Hitchcock brings forward in *The 39 Steps* (1935) and *The Lady Vanishes* (1938) belong to the regimes that would produce the death camps. Now later, the fear of altitude that affects the character Scottie (played by James Stewart) in *Vertigo* (1958) could be an effect of post-traumatic stress common to soldiers—and Stewart was a decorated aviator—returning from the war. So too the gloom that pervades Hitchcock's *The Wrong Man* (1955) could be an aftereffect of the revulsion he felt along with many witnesses of the death camps.[5]

The "mental images" of Hitchcock's cinema oscillate between subjectivation—sentience of the ambient world through fantasy and remembrance—and objectivation—a sifting that separates matter of perception from that of documentation. The spectator is drawn into the films because they can gaze upon skeins of relations in which the characters are unknowingly tangled while feeling (with empathy) at once located in their space and isolated from it. At stake, then, is a double decipherment: characters look into the depth of field in their midst, through the windows of their perceptive faculties, to resolve enigmas while the spectator "reads" various signs and clues, strewn about the surface of the image, that remain invisible to them because they have lost their psychic and geographical bearings. More simply put, real maps, seen in motion or on the screen, beget mental maps that are at once extracted from and superimposed on films and what is recalled of them.

The 39 Steps

An abundant industry of Hitchcock criticism has tended to work less on the silent and British periods and to favor the more opulent features of American post-war facture. In his discussion of the mental image and its knotted relations Deleuze is an exception. He mentions *The 39 Steps* for the symbolic density of the handcuffs the character Hannay (played by Robert Donat) wears from the moment the police (and enemy agents) apprehend him up to the last shot of the film. Leaning over the cadaver of Mr. Memory (played by Wylie Watson), he (unconsciously perhaps) lets one of the rings of the handcuffs dangle from his right wrist, close to the hand of Pamela (played by Madeleine Carroll), the recalcitrant heroine who converted to Hannay's cause and has just helped him break the conspiracy of the 39 steps (Figure 8.1). The symbolic ambiguity of the shape and its virtue as a memory- or relation-image causes Deleuze to link a number of "rings" across three films. He

Figure 8.1 Still from *The 39 Steps* (1935). Source: Hitchcock 1999.

notes that for the *auteur* Hitchcock they are "not an abstraction, but a concrete object bearing diverse relations, or variations of a relation, of a personage with others or himself or herself" (Deleuze 1983: 275). When they command our attention in the final shot, Mr. Memory (an automaton whom the conspiracy "linked" to their operation to remember and then to convey information concerning the compression-ratio for a silent engine to be used in fighter planes) expires in the foreground while a chorus line of dancers mechanically kick their legs in the background.

A welter of relations comes forward: the revelation of political intrigue is meshed with popular entertainment; the conspiracy that began in a music hall now ends in a palladium; Hannay, the tourist or outsider, finds himself the victim of forces or diagrams of power of which he had been unaware; Hannay's half-manacled hand that touches his friend's is supremely ambivalent in that the handcuffs can be appreciated as a grotesque analogue of an engagement ring evincing promise of both bliss and incarceration. Above all, the sight summons many earlier images in which Hannay's ruse and wit—he is a modern Odysseus—bring about a happy end: the medium close-up forces spectators to jog their memory back through the entire film: Hannay eluded capture from the forces of the law by jumping through the window of the police station where he was arraigned. In rapid egress, he suddenly found himself in a parade before he flaked off, again escaping his pursuers, only suddenly to be drawn into a political rally. While deftly biding for time in uttering the required political banalities at the rally he kept his right hand in the pocket of his woollen sport coat, raising his left while using charm to gain a few moments of reprieve. After two conspirators captured and drove him en route to the castle in the Highlands where Professor Jordan had attempted to shoot him, the conspirators handcuff Hannay to the future heroine. He again eluded them, hand-in-hand with his recalcitrant companion; they discovered a rustic inn where they could obtain momentary sanctuary. In registering he forced Pamela to write (with her right hand) a family name for the "newly wed couple" he invented for the occasion. Moments later, his manacled hand grazed her shapely leg as she struggled to remove a wet silk stocking. A skein of possible relations—amorous, political, thespian—gets woven into the words and images of these episodes that are recalled in their "linkage" with the final shot. A cartography of relations is established at the point where a "mental image" of the two hands ends the film.

Thus the film can be plotted and "mapped" with and against its narrative design. Along a similar vector a series of cartographic symbols par excellence determines many of the spatial and psychic tensions of the film. Following the front-credits, the first shot is a mechanical pan to the right that registers, each in isolation, the illuminated majuscules that spell M-U-S-I-C-H-A-L-L. The camera invites a reading of the image as a virtual hieroglyph (m-, mu, mu[te], mus, mus[e], music, mus-i-c [I see], music-hall [all] . . . , and so on) while locating the action of the sequence where Mr. Memory steps on stage. The future hero, a tourist who merely happens to be in the audience, asks a question that remains momentarily unanswered: "What is the distance from Montreal to Winnipeg?" When Mr. Memory gets back to the twice-posed question geography is displaced into the arena of a two-penny opera. When the correct answer comes gunshots and tumult soon follow, setting the throng into confusion and the narrative in motion.

Winnipeg and Montreal in a London music hall? The film ponders the question when it engages a variety of topographies, the first of which is a map shown in extreme close-up. A foreign woman, "Miss Smith" (played by Lucie Mannheim) has latched onto Hannay and asks for sanctuary in his flat, having just fired the gun to evade the agents who were on the verge of apprehending her. A cutaway shot reveals that they are posted in the street below the flat and await her exit. The lady relates her fate in the spare, minimal, whitewashed décor of the flat where she welcomes her host's offer to share a plate of fried haddock. Upon completing her narrative of the conspiracy we ascertain that the gunshots were much like Stendhal's description of politics, in *The Charterhouse of Parma*, "a pistol fired in a concert." Pre-war politics intervene more directly with the map that the attractive counterspy tenders to the future hero at the moment of her death. She staggers toward an armchair where Hannay is sleeping, gasping, "clear out Hannay, they'll get you next," before she expires in his lap. As she collapses a bread knife (the very object he held in his hand while spotting the enemy agents below his window) is shown driven into her back.

The film cuts to a close-up of Hannay's telephone, which began to ring at the instant she collapsed. The camera pulls back (with slight focus pull) to locate the telephone in medium depth into which Hannay moves backwards. He looks at the corpse—but is also looking inward, seemingly lost or aghast in thought—before turning about to notice the men in the street, one of whom is sending a call from the inside of a telephone booth. As he looks out Hannay "sees" the conspiracy through a flashback of the comely agent, now shown behind the mesh of a widow's veil. She utters her words of warning (now at a slower speed and in ominous tones) that she had exchanged with the hero when eating her fish at the kitchen table. He turns, begins to think, and is startled by the crumpled map clenched in her left hand in *rigor mortis*. A medium close-up depicts the moment he extracts the map from her grip. For an instant (Figure 8.2) two left hands, his and hers, touch each other, hers glistening with a marriage ring and a metal bracelet on her wrist, his casting a shadow on the ground adjacent to that of the map. What would appear to be a cutaway shot advancing the narrative is laden with premonitory signs. The bracelet and ring anticipate the handcuff sequences, and the slight contact of the hands is directly related to the last shot of the film when the hero's manacled left hand grazes Pamela's.

Figure 8.2 Still from *The 39 Steps* (1935). Source: Hitchcock 1999.

The scene embodies a cartography of menace. The close-up pulls back into medium view to register the hero gazing at the map, cutting to his hands holding the sheet from a topographic survey of the Tayside region of northern Scotland (Figure 8.3). Roads are shown in broad lines; an ellipse is penciled around a place-name (Figure 8.4). Then is shown, in slightly soft focus around the locale of the town of *Killin*, the detail of "Alt-na-Shellach" (noted near the "Falls of Lochay" and its bridge that will figure in the film) (Figure 8.5). The phone continues to ring so as to confuse the sound of alarm and menace with the sight of the map. The film cuts to Hannay studying the topography before its details, now seen more clearly suggest that he is beginning to look more objectively at the area. The map is quickly overlapped with the image of the woman who now speaks in the manner of a specter, staring through the map while reiterating the ruthlessness of the conspiracy (Figure 8.6). The flashback of the woman speaking through the topographic view, in what might be *a shot unique to early sound cinema*, turns the cartographic object into a "talking memory-map" in which the voice infuses the areas shown (especially "Killin") with anxiety.[6] The map suddenly ventriloquizes by speaking in the voice of its bearer. It signals and even implies the symbolic relation it holds with "Mr. Memory" and his machineries of recall. The extreme close-up of the relief and toponyms transforms conventional signs of the Scottish Highlands into a landscape of fear.[7]

Figure 8.3 Still from *The 39 Steps* (1935). Source: Hitchcock 1999.

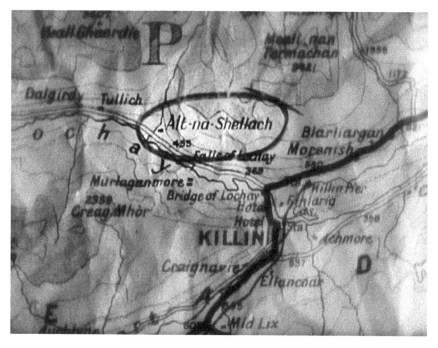

Figure 8.4 Still from *The 39 Steps* (1935). Source: Hitchcock 1999.

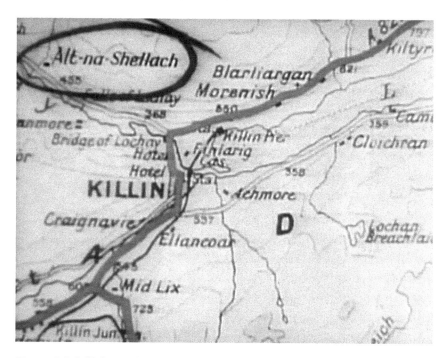

Figure 8.5 Still from *The 39 Steps* (1935). Source: Hitchcock 1999.

Figure 8.6 Still from *The 39 Steps* (1935). Source: Hitchcock 1999.

Contrary to the logic that would use a map to situate the narrative and firmly set spectators in a passive condition in which they let the storyline guide the cognition of the images, the map brings forward the stakes of what it means to read and see in detail, topographically, at once with focus and with a haptic gaze. It prompts the eye to roam about without fixing upon a single shape or following a given set of striations. The projection carries the bonus of shaping the mystery and conspiracy in a frame of escape and pursuit when it becomes, too, a geographic vector that leads the hero north. Hannay's departure from the train station carries the weight of the title—that caused the hero first to quip dismissively, in response to Miss Smith, that "the 39 Steps" might be the name "of a pub"—when close-ups of the feet of the two agents in pursuit are shown in step as they hasten to catch the train departing from the station. Soon, in one of the most anthologized shots of the feature, in a manner complementing the sight of the map and the ringing of the phone in Hannay's flat, a landlady is shown discovering the corpse of Miss Smith. She turns about, looks at the camera and shrieks just as the steaming locomotive exits a tunnel, its whistle blowing and a jet of stream bursting into the air.[8]

The voyage to Scotland can be read as a study of the anxiety of displacement. Hannay, having taken refuge in a compartment where two passengers, the one who would be a virtual old maid and the other a traveling salesman, display clichés of their "Britishness." The train enters and stops at the Edinburgh station where on the soundtrack a boy's voice is heard hawking newspapers. The salesman leans out of the compartment to buy a copy and asks the boy if he "speaks English." The traveler had just displayed his wares, a girdle and a brassiere. The erotic innuendo suggested by the fear that the sight of the underwear elicits (the two Englishmen "repress" their erotic fantasies) shifts onto a touristic plane when the train reaches the Forth Bridge. Desperate when he realizes that his identity is shown on the front page of the newspaper, Hannay hustles from one wagon to the next. To save his skin he enters a compartment in which, to evade the sight of the police, he embraces the woman who will soon appear as Pamela. He then tries to hide in the baggage car before he descends from the train when the police apply an emergency brake. After the hubbub a cutaway shot of the bridge, taken from below and a point to its southeast (as the train might be implied to aim "north by northwest"), seems purely geographical: the famous cantilever structure (5,330 feet long) indicates that the train is at the Queensferry bridge and en route to Aberdeen. In a signature fashion the film turns a monumental structure and a geographical point of reference into a site of anxious anticipation.

A line of divide is met and crossed. At the instant of his escape the hero hides behind a girder that conceals him from the police who have been on his heels. From his ledge he beholds the vertiginous perspective of the waters far below. Here the "mental image" of a network of relations informs the plot and the greater body both of *The 39 Steps* and, broadly, of Hitchcock's cinema. The famous *cantilever* design of the Forth Bridge is exactly what "Midge" (played by June Allyson) deploys in her invention of a new brassiere in *Vertigo*, such that in the later feature the Golden Gate suspension bridge becomes an analogue to the connection between the monument over the Firth of Forth and the British salesman's merchandise. In the film of 1935 the soundtrack intensifies the relation when static noise and a radio emission, fraught with interference, announces that Hannay is "at large" and "is dangerous," when in fact he would be hidden in the maze of beams seen against the evening sky. In all events the bridge and the multiple geographies— spatial, narrative, erotic, political—that cause it to become other than what it seems to be signal that various transgressions are taking place. Historical clichés about antagonisms between the north and south percolate through the image, yet with an intensity so strong that the plot becomes secondary to their presence: which is clear when the agents of the 39 Steps, having apprehended Hannay and Pamela, tell them that they are driving them to Inverness when they are not.[9] Arrested on a stone bridge, a minuscule counterpart to the metal structure in the earlier episode, the couple manages to escape the conspirators thanks to the fact that their car has run into a Scottish icon or cliché *par excellence*: into a herd of sheep whose throng and whose bleats remind everyone that they are in the Highlands and that force recall of the bustle of spectators in the music-hall sequence in London.

At two decisive points in the film a globe—what might be taken to be a counterpart to the topographic map that was cause for Hannay's displacement—displaces the locale into a broader world-picture. After (it is implied) running off the Forth Bridge and getting lost in the Highlands he encounters a grim Scottish crofter (played by John Laurie, also famous for the Scottish character he plays in Michael Powell's *Edge of the World* two years later), to whom he shows the map he had taken with him from London. Hannay gets his bearings, falls into another misadventure in the man's home, but ultimately meets Professor Jordan (played by Godrey Tearle), the leader of the ring of espionage, at the sumptuous castle of Alt-na-Shellach where a pretentiously genteel anniversary party comes to a close. The guests, including the local sheriff (played by Frank Cellier), have just exited. Jordan has shut and locked the doors before he draws a pistol from his pocket and shoots the hero who, we soon learn, is saved by the shield of a bible in the breast pocket of his borrowed overcoat.

The sequence is as remarkable for its geography as its narrative reversals. When the salon is emptied of its guests Hannay is seen alone, taken with his thoughts, no doubt wondering about what he will do and how he will extricate himself from a situation where, on the outside, the police are scouring the landscape and, on the inside, he faces the person whom he suspects to be his mortal enemy. Hannay looks about the room and, it appears, fixes his gaze upon the only doorway from which he can exit. Upon first viewing the shot of the interior seems gratuitous, lackluster, without physical sign of continuity with the preceding shots ... until a mental connection is made. The shot displays in the background a globe that the "haptic" gaze momentarily records. The object that would be synonymous with a well-heeled "bourgeois" or "learned" décor of a gentleman's study begs the viewer (and the hero) to wonder exactly where one really is in the world and, in view of the conspiracy, what the future of the world might be. The globe suggests, too, that the future is at stake, and that the evil professor is at the threshold of keeping the world "in his hands" (albeit with a shortened finger). The sphere subtends a clear ambience of anxiety that reaches within and beyond the screen and movie house of 1935, at the very least insofar as the professor and his tactics are associated with Fascism and National Socialism. The object accrues force as a symbol and a mental image, as it were, of the kind that Gilles Deleuze might have imagined when he coined the terms to describe Hitchcock's cinema and to anchor and confine it in a late phase of the action- and movement-image.

The same globe migrates to the nearby police station in what appears to be the city of Aberdeen. The jovial sheriff, first seen in the salon at Alt-na-Shellach, chides the captured hero and appears to sympathize with his plight. In a two-shot a globe is set between the interlocutors who seek to outwit each other (before the magistrate turns Hannay over to agents assigned to accompany him to London). Once again the future of the world is at stake, even in a place so colonized by its British neighbors that the official feels compelled to insist that his own deeds are the equal of *Scotland Yard*. Is Scotland Yard in Scotland? The globe coyly underscores a sense of displacement—so much so that when Hannay crashes through the window of the precinct in making his escape the globe remains in view, as if to locate the very space from which he eludes his captors. At this juncture (or point of rupture) the globe is affiliated with two forms of law, one belonging to the conspiracy and the other the police, in its pursuit of the hero the latter unwittingly or unconsciously rivaling with the former.

Much as in the sequence in which the topical map of the Tayside region exceeds the narrative it conveys, so also the globe sets the dilemmas of *The 39 Steps* into a greater political arena. Fiction and documentary history are mixed, and so too are physical, psychic, and philosophical geographies. Within the film the maps acquire "locational" functions that cause the viewer to rethink the positions and places from which the fiction can be seen and experienced. The map and globe tend to displace, dislocate, even "deterritorialize" the viewer by eliciting reflection on what it means to look at and to decipher them in a medium of complex visual force. In Hitchcock's work the map can be seen, as Deleuze has shown, as something of a *mental image*. When set in the context of his cartographic lexicon the term allows the map to be appreciated for the way it prompts viewers to discern how cinematic images bear on relations or degrees of consciousness and subjectivity. As a mental image the map engages the process of seeing and deciphering the dilemmas in which a spectator and his or her reflector—a personage such as Hannay—is found. Implied is that the eye studies various degrees of striation and smoothness, or confinement and freedom, and that in its relation with the map it discerns inherited forms or archives (the traditions that inform maps and cinema) that it uses to plot diagrams, that is, new shapes and new itineraries of analysis or even of subjective relations. In *The 39 Steps* the protagonist "works through" the map that locates him in a conspiracy. In its rapport with the world of its time the hero's relation with map becomes informative of the perils of the time in which it was made. As a final note, theory and close study of film show that maps acquire new and unforeseen complexity in the medium and that cinema, in a broad sense, can constantly bear new and powerful forms of cartography.

Notes

1   Discussion of the creative potential of the diagram in the context of resistance is taken up in Conley's (2007: 13) *Cartographic Cinema*. At this point Deleuze is strikingly close to Michel de Certeau's hypotheses about spatial invention. For Certeau space is defined as a discursive practice of place. He or she invents space through a manner of touching, feeling, and even ambulating in areas that would otherwise be under strategic control. Certeau's (1990: 139–93) model derives from work on repressive regimes in South America; its affinities with Foucault's work on mapping and social control are clear.
2   Rancière (2001: 231–5) reads the sequence from the *Histoire(s)* as Godard's attempt to resurrect the image from the ashes of the camps and thus to argue for its potency *nonetheless*, in the wake of its own annihilation.

3   In some dazzling pages Deleuze (1983: 263–6) explicates Peirce's concepts of firstness, secondness, and thirdness through, respectively, Harpo, Chico, and Groucho Marx. After assimilating the analogies the viewer of *Duck Soup*, *A Day at the Races*, or *A Night at the Opera* will never see the films again in the manner of simple comedy. His remarks here bear on *point of view* as it is developed by way of Henry James and Leibniz in *Le Pli: Leibniz et le baroque* (Deleuze 1988: 24–6) by which relation or relativity is discerned from a standpoint that comprehends all possibilities of variation.

4   Following the same line of thought (Deleuze 1988: 25), he notes that new Baroque art tends to be tabular, gridded, like a map of relations on a single surface instead of a window looking out upon a world seen in an imaginary depth of field. What he remarks of Baroque form applies to the way he reads Hitchcock's cinema.

5   In a telling analysis of Deleuze's reading of *Vertigo* Rancière (2001: 154–5) notes that when Deleuze uses formal categories to explain the movement-image he has recourse to thematic or narrative counterparts in order to have Hitchcock anticipate the crisis.

6   In the tradition of British mapping in the early modern age topographic maps of the countryside had been affiliated with a democratic process by which the lands were given to "speak" in the name of the resident population (Helgerson 1992: 33) opposed to the monarchic order of the king. Though the historical past is not directly rehearsed in this moment, the process is nonetheless present so as to make the map a harbinger of menace through its signs of deadly premonition, taken here in the sense of ominous forewarning, in concert with Paul Virilio's (2008) use of the term. Virilio (1982) had already studied cinema as a mode aimed at controlling collective perception of the world.

7   An exchange has taken place. Miss Smith has extended the map to him, and he has accepted it. Following French readings of Hitchcock, Deleuze notes that criminals and those who seek to find them are in a relation of exchange and not of unilateral pursuit. "Rohmer and Chabrol have shown that 'the criminal always commits his crime *for* another,' that the criminal has committed his crime of the innocent person who, as fate has it, is no longer so" (Deleuze 1983: 271). Smith stated as much at the kitchen table when she warned Hannay that he was enveloped in the intrigue before he might know anything of it.

8   Adepts of continuity note that the first shot of the locomotive leaving the station bore the headline "Flying Scotsman" and that the locomotive exiting the tunnel is not the same. The inconsistency reveals how a cinematic ideogram or rebus is built where two different elements are welded to yield a sensation of shock. A complement and a counterpart to Lumière's groundbreaking "Train Entering a Station," similar shots are crafted for the credits of *White Heat* (1949), in which premonition of apocalypse is made clear (Conley 2006: 182–3).

9   "Inverness" figures more broadly on the map of Hitchcock's cinema than through its connotation of narrative ricochet and inversion. It looms in the spectator's memory as the site in *Macbeth* where the ill-fated protagonist had murdered Duncan. By extension, the voyage that goes north also goes to Holinshed's chronicles and more generally to the geography of Shakespeare that richly informs the whole of Hitchcock's cinema.

References

Certeau, M. de (1990) *L'Invention du quotidian 1: Arts de faire*. Paris: Editions Gallimard.

Conley, T. (2006) *Film Hieroglyphs: Ruptures of Classical Cinema*. Minneapolis, MN: University of Minnesota Press.

Conley, T. (2007) *Cartographic Cinema*. Minneapolis, MN: University of Minnesota Press.

Deleuze, G. (1983) *Cinéma 1: L'Image-mouvement*. Paris: Editions de Minuit.

Deleuze, G. (1985) *Cinéma 2: L'Image-temps*. Paris: Editions de Minuit.

Deleuze, G. (1986) *Foucault*. Paris: Editions de Minuit.

Deleuze, G. (1988) *Le Pli: Leibniz et le baroque*. Paris: Editions de Minuit.

Deleuze, G. and Guattari, F. (1980) *Mille plateaux: Capitalisme et schizophrénie 2*. Paris: Editions de Minuit.

Foucault, M. (1975) *Surveiller et punir*. Paris: Editions Gallimard.

Godard, J-L. (1998) *Histoire(s) du cinéma*. Paris: Gallimard/Cahiers du Cinéma.

Helgerson, R. (1992) *Forms of Nationhood: The Elizabethan Writing of England*. Chicago, IL: University of Chicago Press.

Harley, J.B. (2001) *The New Nature of Maps*. Baltimore, MD: The Johns Hopkins University Press.

Hitchcock, A. (1999) *The 39 Steps*. Los Angeles, CA: Delta DVD.

Rancière, J. (2001) *La Fable cinématographique*. Paris: Editions du Seuil.

Virilio, P. (1982) *Logistique de la perception: Guerre et cinéma*. Paris: Editions Gallimard/Cahiers du Cinéma.

Virilio, P. (2008) 'Une anthropologie du pressentiment', *L'Homme* 185–6 (January–June): 87–104.

9 The emotional life of maps and other visual geographies

Jim Craine and Stuart C. Aitken

Introduction

In August 1990, the first Gulf War began between Iraq and a United Nations coalition of forces led by the United States. This moment witnessed a profusion of geographic writing about mapping in the service of empire. Some of these studies documented the relations of cartography to propaganda with a view to unpacking the politics (Pickles 1992), while others revelled with some trepidation in the orthodoxy of so-called 'scientific' mapmaking (Black 1997), and still others sought to demolish that orthodoxy (Wood 1992). With the Gulf War, and the advent of yet more U.S. imperial desire in the Middle East, some geographers took on the new geographic technologies as part of America's superior mapping capabilities to argue the problematic relevance of academic geographers to warfare and imperialism. Neil Smith (1992), perhaps most famously, argued for a moral stance against the Gulf War, and begged geographers to acknowledge the discipline's ties to imperialism. He suggested that GIS developers should accept some degree of responsibility for mapping technologies that are used in the service of warfare. Smith's recriminations are countered by arguments against seemingly divisive polemics and for a more nuanced appreciation of the complex relations between science, civilian mapping, secrecy and classified research (Cloud and Clarke 1999; Goodchild 2006). This is not a new debate. Concern about the use of maps and cartographic technologies as mechanistic tools of war and empire goes back several centuries. The vitriolic debates often miss another form of mapping, one that is used to counter imperial logic and to elicit different, but equally valid, kinds of emotions. Indeed, Guiliana Bruno (2002) creates the beginnings of a revisionist history of mapping that turns the *voyeur* into the *voyageur*, and by so doing moves cartographic emphasis away from *sight*, *site* and what Donna Haraway (1991: 189–91) famously calls the *god trick*, to an appreciation of *motion* and *emotion* and the pleasures of *cartographic embeddedness*. The 'trick' that Haraway draws our attention to is embellished with technical and instrumental discourses that do not necessarily serve local needs. The dilemma of the 'infinite vision', she argues, and the 'promising vision from everywhere and nowhere equally and fully' are problematic representations of 'Science' as 'an illusion'.

With the illusion embraced, our chapter focuses on the emotional life of maps and other visual geographies. It is not an emotive diatribe against geovisualization and GIS. Indeed, Haraway (1991: 181) urges us to embrace new technologies as we seek further insights into the social relations contained therein: 'taking responsibility for the social relations of science and technology means refusing an anti-science metaphysics, a demonology of technology, and so means embracing the skilful task of reconstructing the boundaries of daily life.' And so, we argue for a careful consideration of affective geovisualizations – from maps and other visual media to digital constructs and back again. To do so, we engage the recent literature on non-representational theory, particularly the work of Pierre Lévy and Gilles Deleuze, to question the emotional and affective bases of cartography and to suggest a counter-mapping that celebrates motion and emotion in the service of culture and peace over mechanistic logics in the service of technology and war-mongering.

Affect and the mobility of thought

Let us first explore the role of affect in the mobility of thought and the relation between technique and thinking – in short, understanding how (and why) geographers can focus on the techniques by which visualized information induces affect in the perception of the viewer/consumer. Our question is what affective geovisualization tactics can teach us about the cultural stimuli to thought *and* thoughtlessness, the layering of perception, and the role of affect in perception and thought. We believe that this process of affective geovisualization can inform the reflective techniques we apply to ourselves to stimulate thought and refine sensibility. Through an exploration of the affective processes of visualized spatialities in our increasingly digital environment we can begin to shift the theoretical displacement from the embodiment of the data to embodiment *in the experience of the information*. We can subordinate the function of the computer to the production of affective space. We now correlate affection with the body in such a way that is entirely oppositional to the Deleuzian context that is so prominent in contemporary poststructural geographical theory. Our subordination of the computer exposes the fundamental limitation of the machinic perspective of the neo-Deleuzians: their complete failure to account for the specificity of exteriority, to explain how it bears squarely on the embodied experience of the geographic information. Poststructuralists, especially the neo-Deleuzians, are unable to move past the basic fact that digital information is a new kind of sign system that is no longer mechanical: it produces the separation of the sign and its being (Eisenman 2000). Through affective geovisualization, we argue that the body is now invested with a capacity to convert the digital sign into a somatic experience of affect: Eisenman (1997: 29) makes this very clear:

> Neither the process of the machinic nor any computer program can a priori produce such a trope [a trope in geographical space]. This is what [geographers], weaned on computers, fail to realize. Computers may produce blobs and other self-generated formless aggregations, but these are in and of themselves not any more [geographical] than they are graphic or illustrational. Thus, the [next] step in the process takes a turn away from an aspect of the machine, its possibility of extraction, towards its more arbitrary nature by taking the blurred two-dimensional [image] of superpositions and projecting it into the third dimension.

Thus, rather than creating the production of an affective space in the body, as most neo-Deleuzians do, affective geovisualization functions by creating an affective embodiment that exceeds the spatial bounds of the organic body. Although this interest in the affective process may already be present in geography, we shift the focus from appraising ideological politics, narrative form and cultural message to exploring the geographical relations between narrative flow and specific techniques of delivery – the affective properties of the data and information. We argue that it is indeed these landscapes of geographic data and information that contain the geographical relations between narrative flow and specific techniques of delivery and thus offer new sites of potential geographical analysis through the use of affective geovisualization.

This potential stems from the bodily dimension of affectivity. Bergson (1991), in particular, has located the centrality of affection in perceptual and sensory experience in that there is no perception without affection, meaning that every act or perceiving an object (or image, in this case) at a distance from one's body (or literally, as the potential for the body to act on that object) is necessarily accompanied by an action of the body on itself, a self-affection of the body. Further, Simondon (1989) has expanded this conception by treating affection – or what he calls 'affectivity' – as the mode of sensation that opens embodied experience to that which does not conform to already contracted bodily habits. It does so, Simondon claims, because it mediates between the domain of the 'individual' – the already individuated human being – and the domain that comprises all the processes of individuation (Hanson, 2003). So, where perception appeals to structures already constituted within an individual, Hanson (2003: 108) states that:

> affectivity indicates and comprises the relation between the individualized being and preindividual reality: it is thus to a certain extent heterogeneous in relation to individualized reality, and appears to bring it something

from the exterior, indicating to the individualized being that it is not a complete and closed set of reality.

As the mode of experience in which the individual lives its own thought, affectivity introduces the power of *creativity* into the experiential process. Geographic data, in its cinematic guise for instance, functions to trigger affectivity and operate a transfer of affective power *from* the image to the body thereby, as Lévy postulates, allowing the body to *become* virtual. Instead of a static dimension intrinsic to the image, affectivity thereby becomes the very medium of interface *with* the image. What this means is that affectivity actualizes the potential of the image at the same time it virtualizes the body; the crucial element is neither image nor body alone, but the *dynamic* interaction between them. As our theory of affective geovisualization proposes, if we can allow any visualized geographic data set (such as film or GIS productions) to impact our embodied affectivity directly and more intensely, our geographical engagement with the data – and with visual culture in general – will become more rewarding.

Tender mappings

To put affective geovisualization into the domain of cartography, we begin with a consideration of seventeenth-century counter-mapping and the notion of what we term *tender mapping*. We find that this example is quite appropriate to moving into and beyond imperial cartographies of today and into an understanding of emotional and affective geography. As discussed above, recent cartographic discourse has engaged emotions and other aspects of geographic practice that go beyond representational ways of knowing. Maps are, at base, representations and yet it is not an overstatement to suggest that when they represent space well they also draw us in imaginatively and emotionally. Our understanding of the power of representation is now augmented with non-representational ways of knowing. Tom Conley (2007: 127) notes that a map produced in Madelaine de Scudéry's novel *Clélie* (1654) was very probably drawn in opposition to contemporaneous military cartographies, inaugurated by neo-Cartesian scientists and engineers under seventeenth-century French kings from Henry IV to Louis XIV. French cartographers redrew the defensive lines of the country and designed fortified cities at a time when new siege technologies were changing, with devastating effects, the ways of waging war. Conley goes on to point out that *Clélie* possibly reminded French society of the world of the salon and the space that women had crafted in opposition to the mechanistic world of warfare. Guiliana Bruno (2002: xi) describes Scudéry's novel as a 'sentimental geography'. She goes on to suggest that its map – its 'carte de tendre' – is a celebrated allegory for the female association of desire with space, and an exemplar of the ways in which cartography is inextricably linked with the shaping of female subjectivity. Specifically, the map highlights important passages and mobilities away from lakes of indifference, dangerous seas and *terra incognitae* to favourable villages and towns of tenderness, large hearts, reflection, sympathy and so forth (Figure 9.1).

Jonathan Mallinson (2005) points out that *Clélie* was celebrated throughout Europe in its own time. However, by the eighteenth century it was all but forgotten. Perhaps, then, it served a particular imperial epoch in the way it echoes a romantic view of a pastoral world against the new, hectic (and disease-ridden) urban lifestyles. The novel's *carte de tendre* is a journey through bucolic moralities; it provides a nuanced and clearly categorized representation of different kinds of emotional relationships. At the time, it had its own topical significance and experimental energy, exploring – in speculative or figurative guise – far-reaching questions of freedom and obligation, individual and authority in mid-seventeenth-century France (Mallinson 2005: 396).

The four figures at the bottom right of the map appear to be a man, a boy and two women. Seemingly, the travellers – those seeking enlightenment or simply the consumers – are the boy and the man; and the purveyors of knowledge – the cartographers or the guides, if you will – are the two women. The *Carte du Pays de Tendre*, with its emphasis on such qualities as *reconnaissance* and *estime*, traces a scheme of relationships not just between men and women, but also between kings and subjects, between patrons and writers: as such, it sketches relations between women, men and empire,

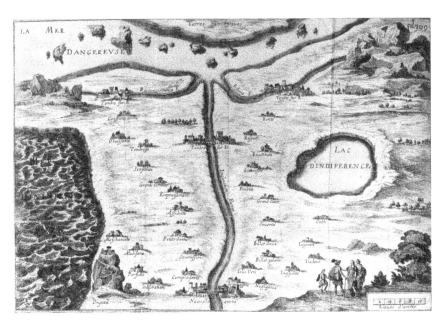

Figure 9.1 Madelaine de Scudéry's *Carte du Pays de Tendre* in *Clélie* (1654).
Source: courtesy of Bibliothéque nationale de France.

authority and subservience, researchers and researched. The *terres inconnues* may not be explored on the *Carte du Pays de Tendre*, but they are recorded: doubt and anxiety have their place alongside confidence and order. These may be the emotions associated with imperial aspirations and mechanistic mappings.

Situated at the cusp of two narrative traditions, Mallinson (2005: 396) points out that *Clélie* transforms and modernizes the heroic world of Scudéry's predecessors, exploring the space between the historical and the contemporary, the literal and the allegorical, the real and the idealized, the personal and the political. It expresses different shades of feeling through the uncertain and ill-defined boundaries that define a journey that divides friendship and love. This is not a new emotional mapping but perhaps emotional relations with space as a forgotten form of mapping. Our argument becomes one that valorizes the intensification of emotional life that is possible through spatial images. Like the work of Scudéry, it seems to us that all represented spaces – from Hollywood films, to propaganda maps to United States Geological Survey (USGS) topographical sheets – are emotionally charged and, perhaps there is a fuller reason to engage with this emotional propensity as we move into an era of possibilities with digitization and the creation of virtual worlds. But first, two brief examples from traditional cartography.

When John Pickles (1992: 193) tackles propaganda maps he is concerned about the ways they are 'discursively embedded within broader contexts of social action and power'. His discussion of texts and hermeneutics is instructive, *and* the maps he provides as examples provide the emotional punch. Pickles (1992: 211) argues, for example, that the use of cartoon octopuses in World War II propaganda cartography highlights the map as an *absence*, which then establishes it as a concrete *presence*. Fine! But when we look at these figures with our students, most comments focus on the demonic faces of Churchill, Uncle Sam and Bolshevism, which head the coiling tentacles that hide the maps below. That is where the emotional power of these representations resides.

In a completely different context, we sit with our students considering a USGS map of the Sawtooth wilderness in Southern California, a map that is ostensibly devoid of propaganda – it is all about contours, hachures, symbols for buildings and roads and such. Certainly, we can deconstruct this like any other map in a bold Harleyian way, but we are also drawn to it imaginatively. This process involves a particularly quirky engagement, pouring over the image to the extent that we may lose a conscious connection to our corporeality. The space between our conscious knowledge of our bodies and the borders of the map merge. Perhaps, for a moment, time disappears. We are lost to the task of imagining what it would be like to be in this place for the first time. We imagine tramping over the hills or along the streets depicted in the map (cf. Aitken and Craine 2006).

We pick up maps of exotic places and indulge our amazement at the contemporaneous heterogeneity of the planet, what Doreen Massey (2005:

15) calls 'spatial delight'. There may even be a geographical imagining that projects knowledge of local weather, waves, tides or of the ways mists curl around mountaintops – knowledge not directly accessible from the map. This is the kind of emotional knowledge that we want to get at with this chapter. At the end of the opening volume of *Clélie*, Scudéry announces that 'sans savoir lui-même ce qu'il voulait faire' (quoted in Mallinson 2005: 396); by so doing she is announcing that the reader, like the hero, like the cartographer, like the writer, is embarked on an epic moral, aesthetic and emotive adventure. In many ways, we argue, this notion of spatial delight is also a notion of methodology, of embedded technique, of visual literacy and the framing of possibilities.

Spatial delight: cinema and the framing of possibilities

We've spent some time elsewhere talking about the relations between cinema as an affective medium and geovisualization as an affective medium (Aitken and Craine 2006, 2008). We've argued that both are about patterns of movement and both are inherently spatial. Our discussion to date begs the question: If cinema is more concerned with engaging emotions than celebrating technology, then why should this not also be the case for research on geovisualization? In broaching this question in this section, we elaborate the emotional content of moving spatial images and make a case for a fuller embracing of non-representational theory by the geovisualization community.

However, a clarification is in order: media is an ambiguous term in and of itself. To Kittler (1999), 'media determine our situation' whereas to McLuhan (1964) 'the medium is the message', an insight into information meaning and technical expression that has important implications for our understanding of media today, especially with the context of geography. Although there is indeed a sort of technological determinism to the creation and consumption of media in all its forms, we wish to address the most fundamental theoretical challenge posed by media to geographers – that of updating geography's study of *all* media. By choice all the media included in this chapter engage a form of *visual* media. Thus, our research most surely finds application to another of geography's central themes – the map, and here we move on to engage film, another of the visual, but nonetheless geographical, forms of media. This is not to prioritize film over any other sort of sensory-based form of media: our work here contains the work we feel best addresses our understanding of the place of media research in our discipline. To that end, our discussion below will seem somewhat cinema-centric only because it is driven by the interest of the authors.

It is important for our purposes here that we first 'place' our theory of affective mapping and geovisualization into the current visual geographic discourse both contextually *and* methodologically. Thorne (2004: 793) makes this apparent:

Visual literacy is an important new skill that geography as a whole needs to embrace for both constructing and deconstructing images. The creation and interpretation of visual images has always been important to geography and is what makes geography unique. It is an exciting time to propose that visual literacy is a common goal of both human and physical geography and that it may act as a common denominator across geography. Common techniques and methodologies are required to both critically understand and to create powerful visual images across the whole discipline of geography.

There is a certain geographic clarity that can be derived from the affective use of geographic information, especially digital data – one of these 'common techniques and methodologies' – because cinema is inherently geovisual and there are emotions behind each visual event. When we 'see' the streets of Los Angeles in any number of films that use Los Angeles as a locale, the image looks extremely *noisy*, but is it possible to look at a Los Angeles street scene without hearing the footsteps and the mingled voices? Taken as an assemblage of individual photographs, each picture is still, without movement or sound, yet, by adding dynamic variables such as sound and light, people begin walking and neon lights start to flash – there is other noise and the image is inhabited by people speaking a language unique to that time and place and space. Clarity about what geovisualization is requires an acknowledgement of the affective *and* the kinetic aspects of cinema – by 'placing' the affective qualities of cinema within the domain of geography we get some specificity of time and space and this clarity furthers our understanding of the nature of geovisualization and protects any insights from simple, mechanistic generalizations. We take as an object of critical analysis geographic data that is often presented as universal and use affective geovisualization to make visible alternative narratives that allow us to *see* more than we already know. To see more than we know is to embrace new knowledges with ambitious, evolving cartographies in their various media forms.

With that in mind, we can consider films to operate as spatial forms – they represent space, place and landscape within a series of frames and these spatial contexts have also shaped the practices of cinema and the meanings contained within films. What we are proposing in this chapter is a weaving of cinema and mapping, a sewing that is so tight that they swap identities. Maps meld into fluid emotions while flickering images are transformed into Cartesian planes. Figure 9.2, for example, shows the map on the wall of the police department at the close of Raoul Walsh's *High Sierra* (1941). Moving images show the police dragnet closing in on Roy 'Mad Dog' Earle (played by Humphrey Bogart), and the map is transformed into an iconic rendering of the gangster's plight. As he seeks independence, the town of Independence is focused in upon, as he moves up into the Sierras to his lonely fate, the town of Lone Pine is depicted on the map. It is a series of 'moving' images,

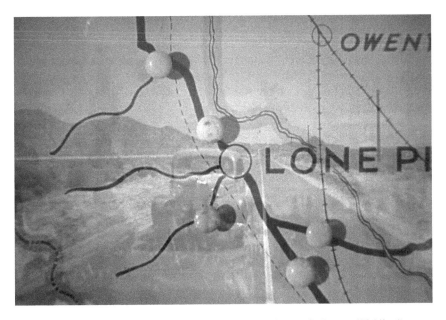

Figure 9.2 Mixing moving image and static map in *High Sierra* (1941). Source:
courtesy of Warner Brothers.

portraying high drama and anxiety; the exhilaration of the hunters and
existential angst of the hunted (see Conley's (2007: 84–91) discussion of
this example).

Film, like mapping, also plays a central role in making social imagery
concrete as part of the 'real' – films have a material effect for those individuals
and social groupings that construct and view them. Therefore, any analysis
of the role of cinema in the discipline of geography involves blurring the
distinction between the real and the imagined. Geographers too often only
consider space as the size of the geographical places and their associated
processes. In other words, space is a macro-environment that exists in space-
time, having complex processes and meaning. But space can also be form
or structure, or pure space, or even space as geometry, or, importantly, it
can be a digital virtual space composed of information. Massey (1992) finds
that space is not static and is constructed out of interrelations creating a
spatiality that is socially constructed. She suggests that space, place and
scale in forms of usage that relate to banal connections (such as flows of
migrants or oil geopolitical maps) can come to be associated with a habituation
and disassociated from their 'full insertion into the political' (Massey 2005:
19). Massey embraces Ernesto Laclau's insistence on the intimate connection
between dislocation and the possibility of politics. For Laclau, spatialization
– for example, mapping space as a Cartesian container of people's activities
– is equivalent to hegemonization. It is a form of control, a framing of

activities into neat bundles that are then amenable to certain forms of analysis from the plotting of migration patterns over decades. Spatial framings such as these are the product of ideological closure, a picture of the dislocated world as somehow coherent (Massey 2005: 25). If we consider space, place and scale as emotive harbingers of the political, as well as an important product of the political, then the lesson from Laclau and Massey is that there must be a dislocation, a freeing from cartography as a framing of possibilities, to a material spatiality that is fluid and open to surprise. This does not in any way soften the concrete interplays between space, community and identity.

Jim Blaut (1999: 511) further adds 'our ideas of pure space are distilled from our space-time experience, by the use of our powers of imagination, or by abstraction.' We argue, therefore, that any analysis must involve a theoretical and methodological approach that engages film as a specific material object existing in space and worth studying as a distinct geographical record within a broader set of practices and discourses.

Becoming virtual: spatial affectivity and visual literacy

The current engagement of geography and cinema is primarily grounded in the work of Gilles Deleuze and his concept of the movement-image – an actualization of the virtual in which images become embodied through an affective process. We argue for a move through (but also with) Deleuze to a more meaningful methodology that permits geographers to more fully engage the digital and virtual geographic data sets not available to Deleuze at the conception of his movement-image discourse. The digital representation was unknown to Deleuze and thus, to accommodate new forms of geographic and cartographic representations, geographers must move instead from the *actual to the virtual*. We believe this move forward, a move beyond (and beside) Deleuze, can be done through a geographic articulation of Pierre Lévy's concept of *affectivity*. Lévyian theory supports a new and importantly different kind of spectorality, one in which the viewer/consumer can enter the places found within the digital spaces of our new technologies. Deleuzian theory does not allow the user's body to enter this space. Lévy, however, privileges the computational power that lies behind digital and virtual technologies, thereby promising an opportunity to more fully comprehend the geographic data coded as an array of iconic images and representations positioned within digital and virtual space. Referent and symbol, subjectivities and icons can now merge as spatialized positions within the coordinated visual worlds of digital space – a space now accessible by our ability to move towards the virtual. This is a new and profound way of mapping the body through its virtual engagement with the technologies and outputs of our discipline, thereby continuing the constitutive technogenesis of geography. We can now develop a technical phenomenology of the body that uses as its modality the virtuality of the human being in its phenomenological affective engagements with geographical mediaspace.

This premise is not new to the social sciences. Mark Hansen (2006) applies the prioritization of the phenomenal body to any engagement with digitally constructed images that mediate the body through technology and thereby constitute embodiment. According to Hansen (2006: 20), this is because digital technologies:

1   expand the scope of bodily (motor) activity; and thereby
2   markedly broaden the domain of the *prepersonal*, the organism-environment coupling operated by our non-conscious, deep embodiment; and thus
3   create a rich, anonymous 'medium' for our enactive co-belonging or 'being-with' one another; which thereby
4   transforms the agency of collective existence from a self-enclosed and primarily cognitive operation to an essentially open, only provisionally bounded, and fundamentally, motor, participation.

Thus, we can then use geovisualization to suggest alternative orderings of knowledge. Geovisualization can take films (or any media) that are presented as natural, universal or true and analyse them so that alternative narratives, based on geography and affect, become visible. We can explore and explain the bond between visual culture and nationalism or gender relations – this will help us understand the motivations of filmmakers to prioritize history by enmeshing the consumer of the image, through the process of affect, in systems of visibility and normalization. With this process of visual literacy, we can also gain some understanding of how dominant classes set up themselves and their heroes as examples to recognize and follow – we can then understand the political interests underlying the production of these cultural representations by using geovisualization to study their transparency: artistic quality matters less than the faithful representation of the achiever or, conversely, the complete subversion of the achiever in an attempt to resist the reinforcement of gender, sexual and racial stereotypes. Geovisualization also promotes the look of the viewer/consumer – understanding the visual production comes first, followed by the perception it guides. We can thus be exposed to the interconnections between private and public, especially private meanings and uses in memories, family histories and the visual tools already in place there and we can also expose the interconnections between the physical and its digital virtual counterpart by in essence becoming virtual. By privileging the movement to the virtual digital dataset, geographers can begin to understand what until now has been considered indiscernible. Films change meaning as the environment changes so the function of visual characteristics in relation to social processes can be the purveyor of a specific relationship to the body. These characteristics can install emotional comfort or distancing, confinement, intimacy or threat, but also, as a cognitive mode of understanding, can provide a 'scientific' method for grasping the complexities of the post-war world. Finally, and most basic, the intertextual relationships between film, as a series of objects or texts, and our different participating senses, the *affectivity* of the image in other words, require this new form of geovisual analysis.

In other words, the viewer – the *consumer* – of the digital image enters that experiential, affective realm between objective seeing and subjective feeling. The user becomes an active participant through a virtual, yet tactile, interface with the image, an image that has the ability to spatialize the body through empirical experience with technology, interfaces that, as Brian Massumi (2002: 192) describes them, 'connect and interfuse different spheres of activity on the same operational plane'. Our bodies become technically inscribed – our sense of vision becomes an instrument to this end – it is a means to a fundamentally tactile interface where vision is bypassed as the essential dimension of the constitution of imagistic knowledge because our bodies are capable of an interface that results in this tactile dialogue with the image.

And it is here that we must depart from the machinistic imperatives of Deleuze's cinematic typology. Geovisualization operates on the premise that the operational success of its productive output is due to its success in confronting the viewer with images that cannot be completed through the action of consumption, actions constrained by the very work imposed on the user in order to solicit knowledge. In this manner, geovisualization functions in a manner similar to Deleuze's cinematic affection-image: like the affection-image, the mechanized interface with the image of the digital representations of geographic data create (if you will) a crisis of sensorimotor logic of the Deleuzian movement-image. Sensation and executed movements overlap in such a confusing fashion that the user is unable to correlate the sensations generated by the data with some appropriate action on the part of the user. However, in direct opposition to Deleuze's typology, affection in conjunction with geovisualized information is *not* a modality of the cinematic image but instead is a sensorial faculty of our embodied singularity. The confusion surrounding our affective embodiment of the information is not the metaphor of Deleuze's cinematic image but is instead the excess of our singular bodies (Hansen 2006). We argue that affective geovisualization extends the logic of affectivity beyond the image by using a post-Deleuzian, post-cinematic interface in which the power of time is constrained by the body's ability to interact within the geovisualized space. This space enables the viewer to utilize the geovisualized information as a simple, goal-oriented temporal exercise to produce an affective experience of the image.

This can be accomplished by the subversion of the dominant Deleuzian *becoming physical* paradigm through Lévy's affective process of becoming virtual. Much of current geographical theory often reduces thinking to cognition, or situates it into an intellectualism or technolectualism that negates its affective sources, somatic entanglements, and effects. But the creative and compositional dimensions of affective geovisualization are essential to freedom of the self and to cultivation of new maps of meaning. One possibility may be a new virtual identity incorporated into the affective sensibility of the individual that remaps the landscape of established constituencies as it struggles to find space.

Filmspace, and visual spaces that employ the same cognitive engagement as filmspace, communicates affective energies to the viewer/consumer, some of which pass below intellectual attention while still influencing emotions, judgements, and actions. We thus argue that purely intellectualist and machinistic models of visual culture should be augmented by a perspective that appreciates the dense interweaving of image, movement, sound and rhythm through the manipulation of affective qualities of the visualized information. We are interested in the relationship between visual technique, feeling, perception and thought and believe that film provides a strategic site of exploration and engagement. Within the filmspace and other visual media are cinematic tropes in which the manipulation of variables such as light, sound, movement and proportion are active agents in constructing affective reactions and are thus open to a deeper geographic analysis through the use of affective geovisualization. For our purposes, we believe these tropes are conducive to providing geographers with a better understanding of how best to engage cinematic space. The cinema and especially its digital counterpart, the DVD (in all its various formats), offer new potentialities for a more comprehensive grasp of the current affective theories advanced by post-Deleuzian media theorists like Lévy, particularly techniques to enable us to mobilize affect and teach us about how such techniques work *outside* the cinematic domain. Affective geovisualization provides a multisensory ontology of the individual that proffers a very different understanding of the affective condition of contemporary technoculture.

Technology and virtual embodiment

In order for that bodily engagement with visualized geographic data to become more rewarding, we must first recognize a valid critique of the model of technics currently in use in geography – that is, its incapacity (or refusal) to come to terms with the correlation of embodiment and technicity. Although the graphical user interfaces (GUI) made their entrance in the 1980s, it was not until the 1990s that computer graphics made a substantial impact on digitized virtual worlds of geography via GIS. Again, the shift was a result of technological advances, but whereas the GUIs paved the way for the user to enter the virtual spaces, the new GIS technologies have been slow to spark the interest of aesthetically inclined people to provide the route into the digital spatialities of the GIS-derived representations. It is striking to see how strong the emphasis has been on the non-aesthetic considerations during the design process of the GIS menu – its GUI in other words, its *doorway* into its virtual environments. Whereas the main issue in the development of a computer application normally is functionality with aesthetical considerations typically left to the end of the design process, we would argue that it should be the other way around with the creation of the visualized dataset. The attraction of the virtual worlds on the viewer/consumer is probably initially that of discovering a new medium to work in, thereby reinforcing the value of Cartwright's (1995) various map metaphors. We can point to at least three aspects of GIS and its mapmaking capabilities that the viewer/consumer might consider important to the process of engaging the digital spaces of the information. First, viewer/consumers are immersed in the design. Second, for viewer/consumers, the design can be seen as a part of the interaction within the digital/virtual world. Finally, conducting research in this new medium leads to working with a new material and, as such, this material can thus reveal possibilities that are unique to GIS and cartographic spatialities. Lévy (1998: 164) puts this into the proper context when he suggests that, '[t]he process of virtualization is only completed with the construction of the object, an object that is independent of the perception and acts of individual subject, an object that can be shared by other subjects.' In this elaboration of virtualization Lévy extracts one significant element of all virtualization processes, the objectivation of a shared context, something contained within the digital multi-universes of the GIS operating environment.

There is no widespread consensual definition of what a virtual world is, even within the context of the GIS environment. The concept is used differently in different contexts. For the sake of clarity and coherence, we offer our own definition to create a common ground for the theoretical and empirical investigation that follows. A virtual world emerges from a technical system that allows a substantial number of people to interact synchronously. The interaction takes place in a sustained environment based on some kind of spatial metaphor. Virtual space is closely connected to non-spatial concepts such as imagination, dreams, memory, religion, etc. The virtual is seen as something 'not-there' rather than something that is there. This leads Lévy (1998: 28) to wonder what happens to Heidegger's being-there as the primary signifier of existence: being in a single, particular graphical virtual world (the GIS environment) where this digital, virtual world of GIS is transformed from a conceptual space (the not-there) to actual places through the being-there. In fact, the physical world must have been just as ephemeral as any virtual world before there was someone there to concretize it through experiences. By being in the world we construct narratives about ourselves that are contextual and fluid. Applied to virtual space, Coyne (1998: 340) advocates a substitution of the language of changed identities with that of narrative construction. Since we all have experience from being in the physical world, the process of understanding a virtual world is more about trying to map our understanding of the physical world onto the virtual world. But at the same time we are always considering new opportunities in the form of aspects of being that are *unlike* rather than *like* the physical world. Lévy (1998) sees this as the actual increasingly becoming virtualized arguing that the virtual must be understood as an historical articulation of the real, fully as actual as any other such articulation, but one connected specifically with computer-mediated communication technologies.

The use of the GIS as a potential database for geographical exploration is rooted within the use of any visual material for research purposes, but particularly the analogue map. Maps have a number of unique qualities that lend themselves to qualitative methodologies: maps can be stored, retrieved, and used over and over again as a kind of permanent record of things seen and see-able. The map, for its part, was fundamental to the rise of Cartesian perspectivism in geography: the map was, as its most basic, a model for conceptualizing human vision based on geometrical optics. We can further elucidate this ideal with the notion that the map locations can be seen as corresponding to a single mathematically definable point from which the visual world could logically be deduced and represented.

Visualizing digitized worlds

Even more facilitating today is a fully digitized GIS environment that, as Wells (1997: 252) points out, makes implicit the shift in the location of final cartographic production from the 'chemical darkroom' to the 'electronic darkroom' of the computer. GIS can be seen as the most current evolution of the cartographic process – it functions as the storage facility for the digital cartographic representation and spatializes the data contained therein. GIS offers the viewer/consumer – and the researcher – the ability to data mine through visual data exploration, a process that Keim *et al.* believe is aimed:

> at integrating the human into the data exploration process, applying human perceptual abilities to the analysis of large data sets available today. The basic idea of visual data exploration is to present the data in some visual form, allowing the user to gain insight into the data, draw conclusions, and *directly interact with the data*.
>
> (2005: 24, emphasis added)

Thus GIS can be seen as a part of the visual data exploration process – one that digitally 'spatializes' the cartographies contained within the digital map. Zook *et al.* (2004: 159) declare that cartographic visualization provides one useful way to envision and begin to analyse the where and how of digital geographies and that, further, spatialization can be considered a subset of information visualization and information retrieval. Indeed, GIS can be perceived as and engaged as a spatialized representation for a virtual environment. Bodum (2005: 392) tells us that 'choosing a suitable data model for the construction of the virtual environment is one of the most important aspects of the modeling. The first stage of the data modeling is a question of choosing the right conceptual model for the virtual environment.' Another aspect of the value of GIS as a depository of spatialized data is Gahegan's (2005: 85) contention that a:

problem with our current geovisualization tools is that they typically are focused only on the data, and not on conceptual structures such as categories or relationships that operate at a higher level of abstraction. Therefore, while they provide good support for data exploration they are less useful for synthesis activities, where concepts are constructed, or for analysis, where concepts are operationalized. To better support the entire science process, we must provide mechanisms that can visualize the connections between the various stages of analysis, and show how concepts relate to data, how models relate to concepts, and so forth.

GIS becomes, then, the mechanism for affective geovisualization – it is a tool that, as Lévy (1998: 94) indicates, 'is more than just an extension of the body; it is the virtualization of an action'. The importance of spatialization and visual data exploration through GIS can be located in their ability to render large amounts of abstract data into more comprehensible and compact *visual* form by generating synthetic spatial structure – and it is these processes of visualization that 'rely on the use of spatial metaphors to represent data that are not necessarily spatial' (Fabrikant 2000: 67) – this will then lead us to a broader understanding on the importance of affectivity in GIS spatialities.

Conclusion: re-imagining cartography

There is today an unprecedented convergence of the analogue map with other, previously distinct imagery, including that produced by GIS, and through computer animation. Digital mapping programs can now directly interface with computer storage and retrieval to the extent that real-time electronic display surfaces only reinforce contemporary society's substantive orientation to visuality (Jakle 2004). And there is a possibility to go beyond the representation, to give an affective push that opens political possibilities.

One possible potential for a broader understanding of affectivity in GIS spatialities is the digital re-imagining of the traditional analogue map so that we may more fully engage it, perhaps enabling an even more emotive rendering that surpasses the affective qualities of the static map. Geographers can now stretch their imaginations around what was unrepresentable – the *terra incognitae*, the monsters and strange beings on the periphery of the known. Perhaps this might soften the starkness of border lines and walls, and produce a rethinking, a re-imagining of the possible from within rather than an imposition of the probably from without and from above. The technology of virtual embodiment suggests the possibility of more vibrant cartographic representations, ones that engage the emotions. The outcome, perhaps, is a completely different kind of cartography, a tender-mapping, an emotional geography of lived experiences. This is the continuation – through digitization, geovisualization and GIS – of a journey out of which the Western infatuation with remotely sensed images, recreated spaces, exotic places, fast motions and fervent emotions has operated for several hundred years but, now, with a twist into the non-representable, the virtual, the real.

References

Aitken, S.C. and Craine, J. (2006) 'Affective geovisualizations', *Directions: A Magazine for GIS Professionals*, 4(1), <www.directionsmag.com/article.php?article_id=2097&trv=1>.

Aitken, S.C. and Craine, J. (2008) 'Into the image and beyond: Visual geographies and GIS', in S. Elwood and M. Cope (eds) *Qualitative GIS: Mixed Methods in Practice and Theory*, London: Sage.

Bergson, H. (1991) *Matter and Memory*, New York: Zone.

Black, J. (1997) *Maps and Politics*, Chicago, IL: University of Chicago Press.

Blaut, J.M. (1999) 'Maps and spaces', *The Professional Geographer*, 51(4): 510–15.

Bodum, L. (2005) 'Modelling virtual environments for geovisualization: A focus on representation', in J. Dykes, A. MacEachren and M-J. Kraak (eds) *Exploring Geovisualization*, Oxford: Elsevier, 3–52.

Bruno, G. (2002) *Atlas of Emotion: Journeys in Art, Architecture and Emotion*, London: Verso.

Cartwright, W. (1995) 'New maps and mapping strategies: Contemporary communications', *SUC Bulletin*, 29: 1–8.

Cloud, J. and Clarke K.C. (1999) 'Through a shutter darkly: the tangled relationship between civilian, military and intelligence remote sensing in the early US space program', in J. Reppy (ed.) *Secrecy and Knowledge Protection*, Ithaca, NY: Cornell University Peace Studies Program.

Conley, T. (2007) *Cartographic Cinema*, Minneapolis, MN: University of Minnesota Press.

Coyne, R. (1998) 'Cyberspace and Heidegger's pragmatics', *Information Technology and People*, 11: 338–50.

Eisenman, P. (1997) 'Processes of the interstitial: Notes on Zaer-Polo's idea of the machinic', *El Croquis*, 83: 21–35.

Eisenman, P. (2000) 'Architecture in the digital age – interview with Peter Eisenman', *Dialogue*, 34: 108–9.

Fabrikant, S. (2000) 'Spatialized browsing in large data archives', *Transactions in GIS*, 4: 65–78.

Gahegan, M. (2005) 'Beyond tools: Visual support for the entire process of GIScience', in J. Dykes, A. MacEachren and M-J. Kraak (eds) *Exploring Geovisualization*, Oxford: Elsevier.

Goodchild, M. (2006) 'Geographic information systems', in S. Aitken and G. Valentine (eds) *Approaches to Human Geography: Philosophies, People and Practices*, London: Sage, 251–62.

Hansen, M.B.N. (2003) *New Philosophy for New Media*, Cambridge, MA: MIT Press.

Hansen, M.B.N. (2006) *Bodies in Code*, London: Routledge.

Haraway, D. (1991) *Simians, Cyborgs and Women: The Reinvention of Nature*, New York: Routledge.

High Sierra (1941) Director, Raoul Walsh. DVD. Warner Home Video, 2003.

Jakle, J. (2004) 'The camera and geographical inquiry', in S. Brunn, S. Cutter and J.W. Harrington (eds) *Geography and Technology*, Boston, MA: Kluwer Academic Publishers, 221–42.

Keim, D., Panse, C. and Sip, M. (2005) 'Information visualization: Scope, techniques and opportunities for geovisualization', in J. Dykes, A. MacEachren and M-J. Kraak (eds) *Exploring Geovisualization*, Oxford: Elsevier, 23–52.

Kittler, F. (1999) *Gramophone, Film, Typewriter*, Stanford, CA: Stanford University Press.

Lévy, P. (1998) *Becoming Virtual: Reality in the Digital Age*, New York: Plenum Trade.

McLuhan, M. (1964) *Understanding Media*, Cambridge, MA: MIT Press.

Mallinson, J. (2005) 'Review of *Clélie: histoire romaine*, by Madeleine De Scudéry: Édition critique par Chantal Morlet-Chantalat. Première partie: 1654. Paris, Champion, 2001. 529 pp. Hb 550 F. Seconde partie: 1655. Paris, Champion, 2002. 529 pp. Troisième partie: 1657. Paris, Champion, 2003. 564 pp. Quatrième partie: 1658. Paris, Champion, 2004. 540 pp.', *French Studies*, 59(3): 396–8.

Massey, D. (1992) 'Politics and space/time', *New Left Review*, 196: 65–84.

Massey, D. (2005) *For Space*, London: Sage.

Massumi, B. (2002) *Parables for the Virtual: Movement, Affect, Sensation*, Durham, NC: Duke University Press.

Pickles, J. (1992) 'Texts, hermeneutics and propaganda maps', in T. Barnes and J. Duncan (eds) *Writing Worlds: Discourse, Text, and Metaphor in the Representation of Landscape*, London: Routledge, 193–230.

Simondon, G. (1989) *L'Individuation Psychique et Collective*, Paris: Aubier.

Smith, N. (1992) 'Real wars, theory wars', *Progress in Human Geography*, 16: 257–71.

Thorne, J. (2004) 'The visual turn in geography', *Antipode*, 36(5): 787–94.

Wells, L. (1997) *Photography: A Critical Introduction*, London: Routledge.

Wood, D. (1992) *The Power of Maps*, London: Routledge.

Zook, M., Dodge, M., Aoyama, Y. and Townsend. A. (2004) 'New digital geographies: information, communication, and place', in S. Brunn, S. Cutter and J.W. Harrington (eds) *Geography and Technology*, Boston, MA: Kluwer Academic Publishers, 155–76.

10 Playing with maps

Chris Perkins

Introduction

Play is increasingly sold as perhaps the most important cultural signifier in people's lives. Indeed, Dovey (2006) argues that play has become the work of postmodern consumer culture, arguing that the commodification of play is central to the ongoing development of global capitalism. From tourist and leisure spaces, public space, to cyberspace and the domestic arena, play allows capital accumulation to progress. Moreover, Eagleton (1987) argues that 'the typical postmodern artefact is playful'. Technological change in the last decade has exacerbated this fixation with aesthetic form instead of function. For example, Manovich (2006: 1) argues that the diffusion and profusion of consumer electronic technologies (such as mobile phones, media players and digital cameras) across society and out of the workplace is associated with modes of interaction such as 'being friendly, playful, pleasurable, aesthetically pleasing, expressive, fashionable, signifying cultural identity, and designed for emotional satisfaction'.

Meanwhile the cultural turn in academic life has been characterized by an interest beyond work and economy: everyday culture matters, style matters, and language becomes a field for play, as well as literal communication (Crang 2000). Writing after the cultural turn has also been increasingly inflected by a retreat from linear narrative, towards irony, intertextual cross-referencing and away from rational notions of progress, towards style and fun. Meanings deployed in visual imagery are no less ambiguous and the visual too can be seen to be subject to, and part of play. So play is on the agenda. Why not use play to rethink maps?

In the 1960s and 1970s cartographic research focused primarily upon communication of information. Play was definitely not part of that agenda. Instead the emphasis was on how map design might be improved scientifically. Research was underpinned by a belief that optimal maps might be produced to meet carefully specified user needs. Universal answers could be discovered through scientific investigation: maps and map use were presumed to exist outside of a social context. This focus inevitably privileged rational scientific ways of knowing the world, whether they involved researching map design, map production, or the psychological processes involved in map reading, in a carefully controlled and experimental fashion. An asocial, artificial and narrowly reductive view of the medium dominated cartography, which took no account of the cultures of mapping (Perkins 2008).

In the 1980s and 1990s, technological change increasingly called into question the fixed format and status of optimal designs of maps and encouraged a profusion of different kinds of mapping. Desktop mapping and GIS gave the general public tools to make their own maps, and thus mapmaking escaped from the dominant control of professional cartographers. GIS allows users to change design specifications and content. Mapping is no longer tied to fixed specifications: users can interact and explore, rather than just employing the image as a final presentation (Rød *et al.* 2001). Morrison (1997) argued that this opening up of mapping technology led to a democratisation of cartography. This democratisation further limits the scope of scientific approaches, at the very time technology has demonstratively altered the significance of mapping. To deal with this radical technological challenge, scientific interest shifted focus towards issues of representation instead of communication, with MacEachren (1995) demonstrating how science might still explain how maps worked as visualisations with complex effects, by fusing cognitive with more semiotic approaches.

In the decade since Morrison's work the Internet has encouraged an explosion in new forms of mapmaking and use and has led to a remarkable sharing of mapping, evidenced throughout this book. The medium becomes much more social and task-oriented, more ubiquitous, ephemeral and mobile. Users and producers are no longer separate. Pervasive technologies offer people the possibilities of putting themselves on their own map, destabilizing the taken-for-granted representational neutrality of the image; new kinds of maps are being made; more people are making maps; more things are being mapped; and mapping is taking place in more contexts than ever before. But understanding maps in terms of cartographic communication or representation still relies upon scientific notions of objectivity and critical distance and underplays the importance of everyday mapping practices.

Different kinds of approaches to cartography emerged in the 1980s that sidestepped the traditional scientific approach: post-structural thought increasingly rejected the possibility of universal explanations and objective measurement and sought more local and contingent insights that welcomed local difference, instead of rejecting it. A social constructivist challenge to traditional map theory emerged, inspired by the work of Cosgrove (1998), Harley (2001) and Wood (1992) and which emphasized mapping as a powerful, social process, in which the image was deployed to normalize and reinforce the position of those with power in society (Crampton 2001). Interests such as militarism, capitalism or racism increasingly formed a focus for researchers, and what came to be termed critical cartography offered new ways of understanding mapping (Crampton and Krygier 2005). These critical approaches have become increasingly diverse and informed by ideas from Barthean semiotics (Wood 1992); Foucauldian power-knowledge (Joyce 2003);

Derridean deconstruction (Harley 1989); post-colonial theory (Sparke 1998) and hermeneutics (Pickles 2004) (see Chapter 1 in this volume for a comparison of approaches).

In this chapter I want to argue that, despite these new critical approaches, an important aspect of the everyday use of mapping has been almost completely ignored in the literature. My argument is that people have always, and perhaps increasingly so, *played* with mapping, instead of simply making or using a map for a specific, instrumental task. Of course there has been some consideration of the role maps might play in games (see Dormann *et al.* 2006 for an exploration of what cybercartography might learn from game design practice). There are also studies of maps as artefacts that subvert, or engage with emotions, notably by Caquard and Dormann (2008) who provide a useful overview of how maps might embody humour. But instead of simply focusing on a single application, or describing playful maps as representations, such as the myriad images brought together on websites such as the Strangemaps blog (http://strangemaps.wordpress.com/), I want to illustrate how a more ludic approach to mapping might be a useful device for understanding how the *process* of mapping, and the map as entity, operate in different social contexts, as a part of life, rather than a tool somehow separate from the culture that they belong to. My argument starts with a consideration of how play as a concept might be understood, and then moves on to consider the contribution this might make towards rethinking maps. Although all maps engender play, some maps are more wrapped up with play than others and a concluding case study illustrates one context where playful mapping is carried out on a day-to-day basis.

Playing the game

It has been argued that mapping and playing have much in common (Lindquist 2001). Both are cultural universals shared by all human societies (Blaut *et al.* 2003). Just as the map can be seen as an entity, with mapping as a process and action, so play is related to the action and process of playing. Both may be understood as a literal activity, but can also serve as metaphors. Both have a contested cultural significance, with often complex meanings charted in everyday uses of the terms in natural language. Table 10.1 explores a few examples for play and playing and shows how richly diverse are the connotations of play.

These uses of play as a noun, verb, adjective, adverb, gerund and past participle reveal three common associations, which Salen and Zimmerman (2003) term game play, ludic activities and being playful. Game play is the formal and focused interaction between players following game rules with some set outcomes; ludic activities are play outside of games and being playful involves the spirit of play being injected into other actions with no particular outcomes. So play can involve being creative, acting something out, activating a process, taking a risky decision, stalling, joking or gambling;

Table 10.1 Playing with language

	Progressive	Fate	Power	Social identity	Imaginative	Self	Frivolous
Playing an instrument	✓			✓	✓	✓	
Playing with toys	✓				✓	✓	
Playing the fool				✓	✓		✓
Playing for time			✓	✓			
Playing the horses		✓	✓				
Playing up				✓		✓	✓
Playing a game	✓				✓		✓
Just playing	✓						✓
Putting into play			✓				
Play around		✓		✓	✓	✓	✓
Fair play			✓	✓			
Wordplay	✓				✓		✓
Foreplay	✓						✓
Play off		✓	✓				
Play with yourself					✓	✓	✓
Play hard to get			✓				
Make a play			✓			✓	
Role play					✓	✓	✓
Played out			✓				
Play ball			✓			✓	
Play of light		✓		✓			
Watching the play				✓		✓	✓
Play of the steering wheel		✓	✓				

it can be a subtle effect, involve fooling someone or being artful. It invariably involves pleasure, is often structured and concerned with sharing meanings in different social contexts. But as a construct it is often easier to define in opposition to other ideas: play is not like work, though it can be a serious business; nor is it synonymous with recreation (for example, when we watch television or listen to music we are not playing).

These cultural connotations illustrate a complex series of rhetorics associated with playing in society, highlighted in subsequent columns in Table 10.1 (Sutton-Smith 1997). Of course many of the terms embody different rhetorics in different contexts. On the one hand there is a widespread and popular impression that play is something somehow wrapped up with childhood, and something we grow out of: the childish process by which we develop so as to be able to work. This *progressive* rhetoric associates play with a phase in an individual's life: child's play, somehow distinct from adult concerns, an educational or improving activity rather than an end in itself. We play *with* toys, with artefacts associated with playful activity, external to us, but embodying playful values. But play also carries connotations of disorder:

something beyond the will, something random, the play of *fate*. The gambler makes a play, which depends on an unknowable and unpredictable random destiny. Nor is all play necessarily childish: it also carries connotations of *power*: almost all sports depend upon competition between different players, which were underpinned in the past (and sometimes still are) by conflict or heroism. Play also carries shared and social connotations, reflected in celebrations, in community values and *social identities*. Many of these social activities are associated with watching play, and behaving as a spectator. A more individual rhetoric is also wrapped up in creative and *imaginative* notions of play, where innovation and individual expression emerge through being playful. Play is clearly also associated with pleasure, with escape, with relaxation and with being an individual: it is a quality associated with the *self* and with individual identity, the eccentric world of the hobby nicely reflects this rhetoric, the converse of serious grown-up concerns such as responsibility, rigour, honour, belief or morality. Play is also about having fun. Certainly many of the connotations of the term are associated with something that is somehow *frivolous*: being playful is not serious. Playing the fool is not something the adult world values. Playing around and playing up illustrate the slightly irresponsible and potentially subversive notions of the term. So play becomes a multi-faceted construct and process, merging progress, fate, power, social identity, the imagination and the self with frivolity.

Yet play has an ambivalent relation to the notion of a game. On the one hand games can be seen as a subset of play, but on the other hand play is a characteristic of games. Caillois (2001) has argued that game motivations can be reduced to four basic tropes, which he terms agôn, alea, mimicry and ilinx. Agôn is reflected in the urge to compete; alea is the play of chance; mimicry is the play of the imagination with its emphasis upon roles and make believe; and ilinx the play of physical sensations. These categories vary according to how structured and rule-bound the activity is: at one extreme are rule-bound games, at the other completely free-form play. But the most complex and complete exploration of the relations between games and play is offered by Salen and Zimmerman (2003), who highlight playing a game as involving goal-oriented, rule-based processes of decision making that are absorbing for players, often strongly social, and uncertain in their outcome, but also often not involving direct material gain or physical harm. They also highlight the voluntary nature of participation, the contest and the play of make-believe, and arrive at a definition of a game as a 'system in which players engage in an artificial conflict, defined by rules, that results in a quantifiable outcome' (Salen and Zimmerman 2003: 80).

Given the complexity of play and its ambiguous relations to games it is small wonder that the activities associated with play have been theorized from many different perspectives. Anthropological understandings of play (e.g. Geertz's 1973 notion of 'deep play') emphasize the symbolic aspects of play as practice, highlighting cultural differences: in Geertz's work for example, Balinese cock fighting represents a dramatization of concerns of social status, but also a reflexive commentary on social hierarchy. These anthropological understandings often contrast strongly with more psychological approaches, which often relate play to developmental states of an individual or child (e.g. Winnicott's 1971 notion of transitional objects). Psychoanalytical studies often overlap with these studies and explore the deeper structural associations of constructs, emphasizing the practicalities of play as a therapy (e.g. Solnit's 1993 exploration of the role of play in child protection and paediatric practice; see also Bingley and Milligan 2007). These contrast strongly with the design-oriented approaches of theorists of game-playing, whose ideas are often driven by the practicalities of industrial production and marketing (e.g. Salen and Zimmerman 2003).

These different discourses around the construct are increasingly coalescing in the multi-disciplinary world of play theory. Building on the classic work of Huizinga's (1955) notion of "homo ludens" an emerging field of study in academia explores playful conceptual depths. However, recent calls for understanding play have questioned the ontological status of the concept and called for understandings that recognise play as embodied, interwoven with place making, affectual and also material (Harker 2005). Thrift (2003) for example, argues strongly that the artefacts associated with new kinds of play, which he terms "supertoys", are themselves far from neutral in the co-construction of playful spaces. So technology and the context in which it is deployed fundamentally alters the experience of play. Research is shifting towards understanding play as culturally and contextually situated (Rambusch 2008). In September 2007 the DIGRA Conference for example, focused on this theme and emphasised that: 'a digital game is an extremely complex aesthetic, social and technological phenomenon. Games are not isolated entities that one can effectively study in vitro. Games are *situated* in culture and society'. Play theory appears to be coalescing around this recognition of plurality, the need for contextual studies, and for an appreciation of cultural diversity.

Given this situated quality it is difficult to reduce play into a single simple set of activities or concepts: on the one hand play seems to be material, an interaction between bodies, environments and artefacts; but it is clearly also increasingly multi-vocal and a strongly contested set of practices that are increasingly mobile, and mutable. For the purpose of my argument here though, I am using a broad definition that takes these difficulties into account and treats play as an open-ended process of investigation in which new worlds are constructed in overlapping worlds of the imagination, cyberspace and reality. How might this notion of play be useful when rethinking mapping?

A playful rethinking of map

It could be argued that all mapping is playful – this metaphorical use of play can be a useful way of understanding many of the tensions revealed in re-imaginings of the power of maps. Many contemporary social constructivist ways of understanding mapping have, in parallel to their scientific alternatives, remained somewhat distanced from mapping practices. Distancing academic research from messy real-world situations, whether in the cause of science, or social critique, risks missing out on important changes. People are making their own maps, and everyday map use is probably more common now than at any time in human history. Almost all of this mapping activity is un-researched. More nuanced and empirically informed real-world studies have emerged recently, informed by ethno-methodology (Brown and Laurier 2005); affect (Kwan 2007); emergence (Kitchin and Dodge 2007); Actor-Network Theory (Perkins 2006); Deleuzian non-representational theory (Crouch and Matless 1996); and holistic performance (Del Casino and Hanna 2005). In this myriad of different ways of thinking about mapping attention shifts onto processes, institutions, social groups, power, interactions between different elements in networks, emotions at play in mapping, the nature of mapping tasks and a concern with practice, instead of focusing on one aspect of how an individual processes combinations of visual symbols on a screen, Web page, or paper sheet. These new ways of rethinking mapping, many of which are described in other chapters in this book, recognize that maps are capable of conveying authority, confirming the subjective, or affirming the cultural taste or place associations of the user, but also recognize that like all iconic devices mapping is only ever able to tell a very partial story about the world, a story that depends upon the situated knowledge of the map reader and his or her culture. Maps are created and deployed in an ongoing and performative process (Perkins 2009) with many similarities to the characteristics of play that are outlined above. Designers play with mapping.

Consider map design . . .

Map design is about beauty and art as well as scientific principles and technology. Keates (1984) for example, explores the many ways in which cartographic design reflects creative and artistic concerns. Ethnographic evidence of designer practice also suggests that regardless of what guidelines or expert systems may say, cartographic technicians explore many options in an intuitive and iterative fashion, deploying technology to "play" with different outcomes, until they reach a solution that satisfies a particular set of criteria (cf. Schienke 2003). They take pleasure in designing a good map. Technological change has increased the possibilities for designers to experiment or play with different looks and functions as solutions can be easily tried out and discarded. The skills associated with making "good" designs are similar to those associated with many aspects of play; with creativity and individual expression important. Playing with map designs can also be fun: consider the visual way that Krygier and Wood (2005) demonstrate these skills, and how much more engaging this is than more conventional written approaches to cartographic design. Design, like play, is about doing, as well as knowing the world, and a more situated approach to map design can flow from technological change.

Instead of a rational process, applying systematic rules to create a single optimal outcome, technology increasingly allows a person making a map to interact with a set of tools, sharing results and negotiating with other stakeholders in an ongoing process that is never really complete, and which only appears complete when the map is called into being by being deployed in an apparently finished form. Mapping emerges gradually like a butterfly from a cocoon, escaping from dead data and transforming into a thing of beauty that flits occasionally into sight/use. For example, Brown and Laurier (2004) show how an ethnographic approach to electronic map design can both deploy the insights of users but also become more reflexive, because of the very technologies that it relies upon to function. The social nature of map design also reflects the social bounding of play. Limiting factors constrain frivolity and the play of the self and depend on context, just as all games depend upon the context they are being played in. For some designers, time-limits dictate style: the media map must hit a publication deadline. For others, more performative qualities of the mapping process are the main concern: the artist deploying mapping is often concerned more with process than final product. For others following a house-style constrains the nature of the play, just as rules in sport discipline and regulate unruly bodily behaviour (Cole, 1993). Designs are deployed and used in playful ways as well.

Consider map use . . .

Ethnographic studies of how people deploy maps in real-world contexts increasingly show that map use is always situated and often social – it is an ongoing and constantly changing process of cross-referencing between many different artefacts and agents. Brown and Laurier (2005) for example, show how different navigational activities involving mapping in a car deploy maps, a driver, a person reading the map, the car, knowledge about where they are, desires about where they are going to, prior knowledge, representations and perceptions of speed, map-reading abilities and so on, together in a process of negotiation. There is an assemblage of actants. The analogies with play are obvious here: a progressive sense of "map skill acquisition" leading to serious map use; playing with maps as a kind of precursor to adult mapping; power, and competition in the mapping – who is in control of the map; notions of identity – preconceptions of who "reads" best; playing as embodied practice; and the active role of the artefact in the process and also the fun of using the map.

Wood (1987) reminds us that there is pleasure in the very idea of mapping. Products evoke sensory responses, which are about more than the mind alone: for example, Harley (1987) shows how a paper copy of a map evokes, embodies and places memories. Graduates responding to questions about their preferences for paper copy or digital mapping valued the hands-on nature of the hard copy map: its tactility and "feel" (cf. Pedersen *et al.* 2005). Blogs reporting on map-reading experiences of hill walkers in the UK also frequently refer to the pleasures of using the map to navigate in the field, as against the functional facility of GPS-based navigation. Designs evoke emotional and bodily responses when they are deployed in task-specific behaviour, and like Harker's (2005) children's playground games, have an affect in the world. They move us – literally and metaphorically: they change the world (cf. Kwan 2007; Craine and Aitken's chapter in this volume). And if all map design and use might be increasingly playful, there are some maps that are even more explicitly part of our society's playful practices.

Consider the contexts for map publication ...

The technological contexts for map publication have themselves encouraged more playful encounters with the map, just as Thrift's (2003) new interactive "supertoys" are changing the nature of play and playful spaces. OpenSource alternatives to mapping are increasingly made by voluntary effort, sourced from crowds, instead of flowing down from hierarchies (Goodchild 2007): part of the world of leisure and pleasure, rather than the directed controlled world of work. During the creation of these citizen-based maps people share information in mashups, and the technology encourages a more playful way of deploying the map. In their analysis of Google Earth for example, Kingsbury and Jones (2009) show how Dionysian impulses may be read into the many alternative and new imaginings of the world, associated with globally available high-resolution imagery. People are playing games with the images, instead of just deploying them to control the world or other people. In the OpenStreetMap (OSM) project, one of the fastest-growing communities of open mapping, participants in the project share mapping instead of owning or controlling the image. Blogs associated with this site frequently poke fun at the serious business of more closed commercial maps (e.g. http://opengeodata.org): for example, the spoof announcement on 1 April 2008 that the OSM Foundation had decided to include fake Easter eggs in its map in order to protect its copyright – playfully poking fun at the practices employed by commercial publishers to deliberately falsify mapping in order to trap those infringing copyright (SteveC 2008). And the day-to-day practice of mapping is fun: mapping parties are held as social events to bring together mappers; people participate because they want to be part of the mapping community instead of as wage slaves (Perkins and Dodge 2008).

Of course a powerful rationale for mapping has always been the drive to leisure and pleasure. With economic development comes a growth in leisure time, and also in sectors of economies catering for playful needs. People in western societies in particular have more time to play, but a characteristic of globalisation has also been the (albeit spatially uneven) emergence of increasing leisure time for newly middle classes. Specialist maps emerged to support this process and the second half of the twentieth century saw huge increases in the number of maps published to meet these needs, and also a diversification in the kinds of maps that were released (Perkins and Parry 1996): tourist maps market a place but also help us to imagine our holiday; specialist topographic or walking maps help us to climb a hill; cycling maps focus on possible routes; orienteering maps allow the sport to take place and so on.

A detailed focus on a single leisure pursuit reveals a huge diversity of different kinds of mapping catering for very different parts of play. Golf is a sport that pits an individual against a landscape, in a strongly regulated game, where play is a carefully articulated series of individual performances, in subtly different places: every golf course is unique (Klein 1999). Golf mapping has emerged in particular since the 1960s to represent different aspects of these landscapes and practices, but is deployed in a very wide variety of different contexts underpinning the playing of the sport. Maps place and locate courses; they allow courses to be sited and markets to be assessed; help architects to play with different routings and design holes; facilitate course management; measure distances on the course for players; play an important role in promoting the course; help to control individual golfing performance; but also stimulate the memory of events and places. It has been argued that these deceptively simple graphics are deployed as an important and embodied part of the game, rather than simply serving as representations of golfing landscapes (Perkins 2006). It is play that forms the rationale for all of this mapping.

The following case study focuses on the creation of one particular set of golfing maps exploring the roles they play in online computer gaming and relating this playing with maps in particular to Sutton-Smith's (1997) rhetorics of play, and to more recent affectual and situated theorising about mapping and playing. Playing golf video games is particularly appropriate for my argument here. The game serves as an artefact, with a material presence, akin to Thrift's (2003) "supertoys": without the external material presence of the software, and its enacted and embodied visuality on the screen, play could not take place. But the golf video game also imitates the real world. At the same time the medium allows mapping practice in the game to be documented, in a way that is difficult in the real world of golf mapping. Also the *act* of mapping in the video game is itself carried out as a central part of playing the simulated game. The players are literally playing with the maps.

Playing with golf maps

By 2008 computer gaming was one of the most popular leisure pursuits in Western Europe and North America, a multi-billion-pound industry fast catching up on the film and music industries in terms of revenue (Bangeman 2008). In April of that year sales of Take Two Interactive Software's *Grand Theft Auto IV* topped $500 million in a single week, making it one of the most lucrative entertainment events ever (Paul 2008). These games may be grouped into different kinds according to the interplay of what Brand and Knight (2005) describe as ludological and narratological characteristics. Ludological factors include topography; pace, representation and nature of time; player structure; control; mutability; determinism; savability; rules and strategic objectives. Factors relating to the story of the game include formality; the nature of the narrative model; architecture; degree of player influence; temporal setting; story order; range, depth and fluctuation of story information. Different genres have emerged and are strongly marketed by the industry, notably action, adventure, educational, driving, role-playing, simulation, sports and strategy games. These genres are played on an ever-changing series of different platforms, with different functionalities and user demographies: a strongly competitive market sets the context in which this play is able to take place.

Golfing games on the computer have been readily available since the 1978 release of *Computer Golf!* on the Odyssey 2 platform. In May 2008 the most comprehensive game documentation and review project website Moby Games listed a total of 244 different golf games, out of a total of 2,612 computer and video games in the sports genre: the second-highest number behind soccer. This total includes significantly different games, with varying ludological and narrative characteristics. The majority of these games allow players to compete against the course, or against other players, reflecting Caillois's (2001) notion of agôn. Game play simulates playing on a course that appears on the screen in front of you, and in which some form of input device becomes the club, with which you simulate the golf swing and in so doing strike the ball so that it moves across the virtual layout of the course on screen.

However, the player is never completely in control. As in other sports games a probabilistic engine partly determines the outcome of shots: Sutton-Smith's (1997) play of chance strongly influences computer-golf outcomes. Nevertheless it has been demonstrated that the progressive rhetoric of playing with simple golfing video games can lead to improved performance in the real game. In a systematic evaluation of performance Fery and Ponserre (2001: 1035) concluded that 'if the user is engaged in a learning skill strategy, golf video games seem to be useful in sport skills acquisition'.

Recent innovations such as *Real Play Golf* and the Nintendo Wii have introduced bodily action into player control, leading to a more immersive feeling, and presumably enhanced gameplay training potential. Even the less radical *Tiger Woods PGA Tour 08* allows players to superimpose a photo of their own face onto game characters. But these immersive trends also reveal the limits of a purely representational view of the mapped course: instead the course is best interpreted as part of the game, which is made through the bodily practices of the players.

Almost all computer and console-based golf games depend upon a mapping of the course against which the game is played. Indeed Aarseth (2001:

15–16) argues that this concern with space is what most characterises the computer game as a cultural form and deploys a Lefebvrian argument to illustrate his case: 'computer games are both representations of space (a formal system of relations) and representational spaces (symbolic imagery with primarily aesthetic purpose).'

In many computer-based war-games the map plays a function that is much greater than simply representing certain features in a particular style: as Salen and Zimmerman (2003: 444) put it 'the meaning ... arises not just from its geographic or pictorial features, the meaning derives from the role the map plays in the larger game experience.' This wider contextual role for mapping is certainly also valid in the case of golf games.

In character-based role playing games Dormann *et al.* (2006: 54–5) characterise two main kinds of mapping, with rather different roles for the player: the fantasy world map, whose 'main purpose is to give a sense of place in the game and increase realism'; and the functional navigation map placing players that emerges gradually and progressively during exploration of the world. The game player is simultaneously inside the virtual world, but also inside the map, so an appreciation of different perspectives becomes more possible – an ambiguity also evident in golf games.

Mapping in golfing games is also both conceptual and associative. Just as mapping the real game of golf fulfils different roles, so these golfing games emphasise different uses of mapping. Many of the games are highly stylised, and only very loosely related to the real game of golf. The console market in particular emphasises game play, competition and players, instead of the landscape and course on which the game is played, and 'fantasy golf' abounds in titles such as *Mario Golf*, *Hotshots Golf Fore!*, or the over-the-top, violent and comedic *Outlaw Golf*. The cartoon-like characters, and extreme courses in these games actively subverts and parodies the rather staid, conservative and regulated play of the real game, evoking the subversive qualities of play. Here mapping is a backdrop to the action – and little attention is paid to where the action takes place.

A second kind of golf game focuses on simulations such as *Sid Meier's Sim Golf* (see Figure 10.1e) that emphasise playing at managing the course. These simulations stress the planning of the whole course, rather than playing on courses created by other people. The aim is to design an attractive facility, so as to attract customers to the course. Maps are the interface through which the course is made, rather than just serving as the ground on which the game is played. The level of hole-detail in this kind of game is limited: mapping is a device to illustrate animated success in attracting revenue. Realistic depiction of a golfing landscape is not the aim – a stylized depiction and isometric view depicts the emerging golfing landscape: here the user, as part of the game, makes maps.

The third and most complex kind of golf game uses course representations as a central feature of the game, with almost photo-realistic depictions of course scenery. These attempts to simulate the real golf course, with game play under user control against a mapped or photo-like backdrop, have been published over the thirty-year history of the development of golf gaming, and often in long-running series of games. Important series include the *Links* franchise (see Figure 10.1f), released over the period from 1990 until 2003 by Microsoft; and franchises from Electronic Arts, *the PGA Tour* series (1990–8) and the current market leader *Tiger Woods PGA Tour Golf* (see Figure 10.1h). Different course backgrounds have been supplied with games since the early days of the genre, emulating classic real golf courses such as Augusta, Georgia or St Andrews. Users play on courses supplied with the software, or download or purchase new courses. The elements of the course are mapped in a detail that strives to emulate the real-world experience of playing a course. Rough is clearly distinguishable, greens are shown in higher resolutions than fairways, different tees are carefully distinguished and distances precisely mapped out.

However, new mapping can also be created by the players themselves in many of these games, and in some ways this user capability makes a fourth

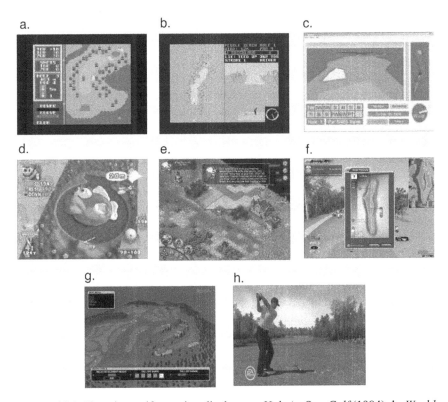

Figure 10.1 Changing golf-mapping displays. a. *Hole in One Golf* (1984); b. *World Tour Golf* (1986); c. *Fuji Golf* (1991); d. *Mario Golf N64* (1999); e. *Sid Meier's Sim Golf* (2002); f. *Links* (2003); g. *Custom Play Golf* (2005); h. *Tiger Woods PGA Tour 08* (2008). Source: author screenshots.

kind of golfing game. The first course-editing suite to be made available with a game was released in *World Tour Golf* in 1986 from Electronic Arts (see Figure 10.1b). The *Links series* encouraged users to share courses, a trend continued in *Tiger Woods PGA Golf Tour*. Even *Sid Meier's Sim Golf* allows users to make and share their own courses. Substantial numbers of user-designed courses are now available over the Web.[1] Other products trade on the novelty of being able to play on a new course at every game – for example, with *InfiniTee Golf*. This more creative rationale for play is most developed in *CustomPlay Golf* (see Figure 10.1g), where software sophistication overlaps with that deployed in the real world of golf-course architecture, and where making the course is marketed as being as important as playing on it.

The functionality and design of mapping interfaces deployed in these games have changed over time. Figure 10.1 shows something of the changing aesthetics involved in this mapping process. Improvements in screen resolution and processor speed have allowed an increasingly realistic depiction of landscape to take place. Mapping has become much more complex, with increasingly sophisticated landscape depiction and often control over textures, models, sound effects and the ways these relate to game play. Early games offered only limited vertical views onto the golfing landscape. Split-screen vertical and third-person perspective views of the course appeared in the mid 1980s in *World Tour Golf* (see Figure 10.1b). Current games support multiple views, zooming in on characters and parts of the course, deploying default or user control. Three-dimensional displays allow panning and zooming around objects. There has been a shift towards multi-window designs, incorporating zoomable maps, where the viewing angle can be changed. Users increasingly are able to control these map views, selecting whether to see the course from the player, audience, birds eye, or isometric perspective. The sophistication of game play has been greatly improved by the development of smoother animation and better modelling of the physics of shots and golf ball flight.

There has, as elsewhere in the gaming world, been a shift towards multi-player gaming, onto multi-site competition and towards multiple platforms for the same game. *Tiger Woods PGA Tour 08*, for example, is available for PC, Playstation1 and 2, Wii, DS, Xbox 360, and PSP. The Web now facilitates new kinds of playing: online competitions and tours mirror the real world of golf competition. For example, *Shot-Online*, developed by OnNet Co., Ltd, and released in 2006, is a hybrid, massively multiplayer online golfing simulation game, incorporating elements of role playing (such as training and mini-competitions), enhanced by a proprietary physics engine for golf-ball dynamics and interaction with eight different courses. But just as wider notions of play are increasingly being studied as social and situated (Rambusch 2008), so has playing with golf maps become increasingly social. A more detailed consideration of the practices around a single golfing game reveal the importance of moving beyond function and representation.

Playing practices with Microsoft Links

The *Microsoft Links* series of games illustrates some of the most sophisticated kinds of interaction and mapping capability, and also shows the importance of a situated understanding of playing with maps. *Links 2003* was marketed with six built-in courses and an expansion pack comprising a further twenty courses. There are now forty-seven courses sold from the official Links Country Club website (http://linkscountryclub.com). The package also included a design engine: the Arnold Palmer Course Design (APDC) module (Figure 10.2). This allowed users to create their own courses with sufficient flexibility to mimic real-world golf course architecture. Greens, rough, bunkers and fairway could be customized, elevations altered and the detailed topography of courses manipulated. The 'look' of the course could be changed as well. Design has been decentred, just as in the world of open-source

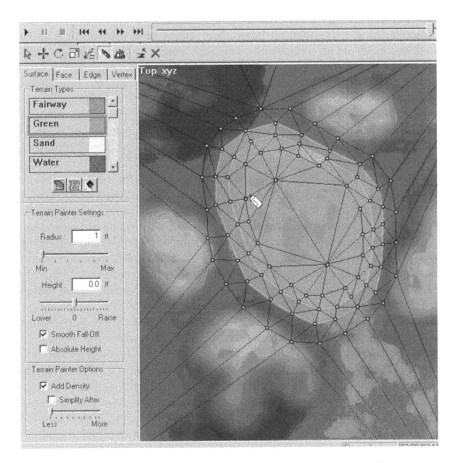

Figure 10.2 Tools for designing golf maps online: APDC mapping interface. Source: author screenshot.

mapping: the technology allows multiple roles to be played out by the same person, instead of separating 'skilled' producers from 'unskilled' users. The users also create the code, with significant libraries created in the community, enhancing the original software.

The Web-served environment in which it was released allowed a substantial community of users to develop, who exploited the flexibility of the program to make an impressive number of new courses. The practices of game playing themselves made new playful spaces. Links Corner (www. linkscorner.org) is the best-developed community website associated with this package, reportedly with over 15,000 members. It offers free download of the APCD software as well as a place to share community-designed courses. By 2008 there were 967 APCD courses available as free downloads on the site, which continues to flourish despite the lack of upgrade of the playing software itself. New courses are released for testing in a collaborative and iterative approach. The competitive ethos of sports gaming ironically depends strongly upon a co-operative sharing of information. The site also offers specialist forums in which the community shares ideas about design, about courses and to set up tournaments. By 2008 there were approaching 20,000 posts on these forums.

Just as play replicates and imitates elements of the real, but also allows the imagination to make new contexts (Sutton-Smith 1997), so *Links* overlaps with the real world of golf, but differs from it (see Figure 10.3). Course designers classify courses when they are uploaded into one of the following categories: desert, heathland, links, mountain, ocean, parkland, tropical or woodland, just as in the real world these broad classifications are used in the marketing of courses. Designers allocate a par to the course, calculate its length from the back tees, and provide a brief description of the course for beta testers of the software. A 'difficulty index' is created for each course that is played, which parallels the real-world USGA slope rating system. Tournaments online also imitate real tournaments. Membership of online *Links* forums overlaps with membership of online forums where real golf course architecture is critiqued, notably Golfcourseatlas.com, and the designers of some of the gaming software also offer turnkey real-world course design solutions: just as real-world course architects are venerated in their community, so are star online designers such as Mike Jones lauded online. Their online courses must be classified as real, fictional, imaginary, a demo or a beta test: game play, however, allows more than real elements to be incorporated into designs; surreal contexts offering extreme golf are possible. Game play literally depends upon "make believe" as well as on imitative qualities, and to be successful in *Links* depends upon being creative.

This collaborative mapping exercise can appear daunting to novices and is almost certainly a minority pursuit among game players. The community has its own specialists and experts in very specific aspects of the course/map design process. There is a large playing audience who download created courses/maps, and a rather smaller creative class who develop the necessary

Figure 10.3 Links 2003 12th hole, Augusta, Georgia, designed by Mike Jones 2001, compared with a photograph of the real 12th at Augusta, Georgia. Source: author screenshot and http://linkscorner.org.

software skills. To encourage a wider and more active participation in course-making the site also offers numerous tutorials, for example, in AutoCad and Photoshop, where skills may be shared among users. Sutton-Smith's (1997) *progressive* rhetoric of learning through play is clearly important here. The course mapping process is, however, never really completed: one map feeds in to another, as designers learn from each other's practice. Courses pass from one designer to another, and are frequently amended, with mapping repeatedly called into being and changed, in the ontogenetic process documented by Kitchin and Dodge (2007).

The *power* of play is also evident in *Links*. The game itself is strongly competitive, the goal is after all to win, but *Links* Corner also includes a detailed reviewing section, in which new designs are critiqued for background, impression, playability, visuals and technique. The goal here is to produce aesthetically satisfying courses and maps with appropriate sound, scenery and game play. The community polices members, powerfully approving or disapproving of playful practice. Officially sanctioned panels of reviewers and careful peer evaluation of designs dictate acceptable behaviour online and rank the best courses: it is the most-approved mapped courses that are most downloaded, and ironically the most famous real courses are also the most popular *Links* course. Figure 10.4 illustrates aspects of the peer-review process for one of the most popular online creations.

Postings on *Links* Forum also reveal the social significance of participation reflecting the rhetorical play of *social identity* (Sutton-Smith 1997). Participation is strongly gendered – membership of the *Links* Forum is even more male than the predominantly male world of real golf. Membership is international, but mainly North American, again reflecting the dominance of North America in real golf-course architecture and in the sport in general. The social context of the playing with maps is strongly interactive. Participants share knowledge and learn mapping skills from one another. Designing a

Closing comments of review by Chris Gormley:

Like I stated above this is probably one of the best real course recreations I have seen for Links. Chuck has taken his designing to a new level and we the Links Community get to enjoy his talents once again. I remember purchasing the Access version of Pebble Beach and thinking it was one of the best courses we had in the community at the time. After playing some rounds on Chuck's version of Pebble Beach I have the same feeling, this is one of the must have courses that we have available to us. There is nothing about this course that should deter a player from downloading this one and keeping it on the hard drive for a very long time. I highly recommended Pebble Beach 2004. *"I am writing this review from the viewpoint of not only a player but as a Tour Director. Although I am not a designer I understand the work that goes into a design as I have beta tested many courses and my review is simply how I as a player/Tour Director would view this course."*

	Votes cast		
Ace	▪	88 %	190
Eagle	▬	11 %	24
Birdie	▪	0 %	1
Par	▪	0 %	1
Bogey or worse	▪	0 %	0

Figure 10.4 Reviewing golf-map design: *Links 2003, Pebble Beach 2004* designed by Chuck Clark. Source: http://linkscorner.org/courses/course.php?crz=1592.

course is a time-consuming process and group membership is strongly valued online: playing with these maps is a time-consuming hobby reflecting Sutton-Smith's (1997) rhetoric of play and the self.

The affectual qualities (Harker 2005) of mapping practices in *Links* are also clearly evident. Many of these might be seen by outsiders as being associated with Sutton-Smith's (1997) rhetoric of *frivolous* play. Participation is after all about having fun. But examination of the forum reveals more complex emotions at play. Emoticons are frequently deployed in the *Links* postings, conveying feelings about the process, which reflect many different emotions: frustration at poor play or an inability to grasp a difficult design concept; a sense of achievement in a tournament victory; a sense of belonging to a like-minded community and so on. It is these emotions that are driving the making of new courses and spaces for play. Avatar identities in the Forum also illustrate real-world aspirations: participants are passionately involved in the creative process of making new maps and courses and in playing and competing on these maps. They take pride in their creations and in the success of their courses in the eyes of the community. They may be playing as amateurs, and for fun, but participation is a serious business, often of central importance for the identity and practices of people involved.

Conclusions

The zeitgeist of the emerging discipline of game studies is shifting from a representational and cognitive focus, to a much more nuanced and performative understanding of computer games as a form of situated play (cf. Rambusch 2008). This chapter has explored an approach to mapping, focusing on the metaphor and practice of play, as an open-ended process of investigation in which new worlds are constructed in the imagination, in cyberspace or in reality, which reflects this changing conception of play. It has shown how theorizing mapping has largely so far ignored play: but also that online mapping tools, collaborative mapping and new contexts have enabled new ways of thinking about mapping. I have argued that there are clear analogies between new conceptions of play and these new mapping practices. Thinking about mapping has moved on to reflect the complex and social ways in which representations are deployed in different cultural practices, and the process of designing mapping is increasingly analogous to playing. The ways maps are deployed and used also seems to parallel play as a social practice. Some contexts in which mapping is called into being are particularly playful and an examination of one of these contexts shows how the detail of everyday mapping practice is played out.

This chapter has also charted experiences of collaborative mapping, in contexts where designs are shared and critiqued, where mapping *for* play, is itself played *with*. Case studies of online mapping in computer and console-based golfing simulations have been employed to show the potential of playing with map design, playing with, and on, the map. The relations, interactions and practices, rather than the end products, seem to be what matters in this process. The performance of mapping reveals that there is still everything to play for in this kind of rethinking of the mapping process and that mapping and playing are cultural practices in which the politics of affect are played out.

Note

1   By May 2008 there were: 2,586 *Tiger Woods PGA Tour* courses, www.coursedownload.com; 1,521 *Sid Meier's Sim Golf* courses, www.simgolf.ea.com; 967 *Links 2001* and *Links 2003* courses, www.linkscorner.com.

References

Aarseth, E. (2001) 'Allegories of space. The question of spatiality in computer games', in M. Eskelinen and R. Koskimaa (eds) *Cybertext Yearbook 2000*, Saarijrvi, Finland: Publications of The Research Centre for Contemporary Culture, University of Jyvskyl, 152–71.

Bangeman, E. (2008) *Growth Of Gaming In 2007 Far Outpaces Movies, Music*, <http://arstechnica.com/news.ars/post/20080124-growth-of-gaming-in-2007-far-outpaces-movies-music.html>.

Bingley, A.F. and Milligan, C. (2007) 'Sandplay, clay and sticks: Multi-sensory research methods to explore the long-term influence of childhood play experience on mental well-being', *Children's Geographies*, 5(3): 283–96.

Blaut, J.M., Stea, D., Spencer, C. and Blades, M. (2003) 'Mapping as a cultural and cognitive universal', *Annals of the Association of American Geographers*, 93(1): 165–85.

Brand, J.E. and Knight, S.J. (2005) *The Narrative and Ludic Nexus in Computer Games: Diverse Worlds II*, <www.digra2007.jp/>.

Brown, B. and Laurier, E. (2004) 'Designing electronic maps: An ethnographic approach', in L. Meng, A. Zipf and T. Reichenbacher (eds) *Map-based Mobile Services – Theories, Methods and Implementations*, Berlin: Springer Verlag.

Brown, B. and Laurier, E. (2005) 'Maps and journeying: An ethnomethodological approach', *Cartographica*, 40(3): 17–33.

Caillois, R. (2001) *Man, Play and Games*, Chicago, IL: University of Illinois Press.

Caquard, S. and Dormann, C. (2008) 'Humorous maps: Explorations of an alternative cartography', *Cartography and Geographic Information Science*, 35(1): 51–64.

Cole, C.L. (1993) 'Resisting the canon: Feminist cultural studies, sport and technologies of the body', *Journal of Sport & Social Issues*, 17(2): 77–97.

Cosgrove, D. (1998) *Mappings*, London: Reaktion Books.

Crampton, J.W. (2001) 'Maps as social constructions: Power, communication and visualization', *Progress in Human Geography*, 25: 235–52.

Crampton, J.W. and Krygier, J. (2005) 'An introduction to critical cartography', *ACME: An International E-Journal for Critical Geographies*, 4(1): 11–33.

Crang, P. (2000) 'Cultural turn', in R. Johnston, D. Gregory, G. Pratt and M. Watts (eds) *The Dictionary of Human Geography*, 4th edn, Oxford: Blackwell, 141–3.

Crouch, D. and Matless, D. (1996) 'Refiguring geography: The parish map project of Common Ground', *Transactions of the Institute of British Geographers*, 21: 236–55.

Del Casino, V. and Hanna, S.P. (2005) 'Beyond the "binaries": A methodological intervention for interrogating maps as representational practices', *ACME: An International E-Journal for Critical Geographies*, 4(1): 34–56.

DIGRA (2007) *Digital Games Research Association 2007 Conference Overview*, <www.digra2007.jp/Overview.html>.

Dormann, C., Woods, B., Cacquard, S. and Biddle, R. (2006) 'Cybercartography as a role playing game: From multiple perspectives to critical thinking', *Cartographica*, 41(1): 47–58.

Dovey, J. (2006) 'How do you play? Identity, technology and ludic culture', *Digital Creativity*, 17(3): 135–9.

Eagleton, T. (1987) 'Awakening from modernity', *Times Literary Supplement*, 20 February.

Fery, Y-A. and Ponserre, S. (2001) 'Enhancing the control of force in putting by video game training', *Ergonomics*, 44(12): 1025–37.

Geertz, C. (1973) *The Interpretation of Cultures*, New York: Basic Books.

Goodchild, M. (2007) 'Citizens as sensors: The world of volunteered geography', *GeoJournal*, 69(4): 211–21.

Harker, C. (2005) 'Playing and affective time-spaces', *Children's Geographies*, 3(1): 47–62.

Harley, J.B. (1987) 'The map as biography: Thoughts on Ordnance Survey map, six-inch sheet Devonshire CIX, SE, Newton Abbott', *Map Collector*, 41: 18–20.

Harley, J.B. (1989) 'Deconstructing the map', *Cartographica*, 26(2): 1–20.

Harley, J.B. (2001) *The New Nature of Maps*, Baltimore, MD: Johns Hopkins University Press.

Huizinga, J. (1955) *Homo Ludens: A Study of the Play-Element in Culture*, Boston, MA: Beacon Press.

Joyce, P. (2003) *The Rule of Freedom*, London: Verso.

Keates, J.S. (1984) 'The cartographic art', *Cartographica*, 21(1): 37–43.

Kingsbury, P. and Jones, J.P. (2009) 'Beyond Apollo and Adorno: Dionysus and Walter Benjamin on Google Earth', *Geoforum*, forthcoming.

Kitchin, R. and Dodge, M. (2007) 'Rethinking maps', *Progress in Human Geography*, 31(3): 331–44.

Klein, B.S. (1999) 'Cultural links: An international political economy of golf course landscapes', in R. Martin and T. Miller (eds) *Sportcult*, Minneapolis, MN: University of Minnesota Press.

Krygier, J. and Wood, D. (2005) *Making Maps: A Visual Guide to Making Maps for GIS*, New York: Guilford Press.

Kwan, M.P. (2007) 'Affecting geospatial technologies: Toward a feminist politics of emotion', *The Professional Geographer*, 59(1): 22–34.

Lindquist, G. (2001) 'Elusive play and its relations to power', *Focaal – European Journal of Anthropology*, 37: 13–23.

MacEachren, A. (1995) *How Maps Work*, New York: Guilford Press.

Manovich, L. (2006) 'Interaction as an aesthetic event', *Receiver*, 17: 1–7, <www.vodafone.com/flash/receiver/17/articles/pdf/17_09.pdf>.

Moby Games (2008) *Game Browser*, <www.mobygames.com/browse/games/x,5/golf/>.

Morrison, J.L. (1997) 'Topographic mapping in the twenty-first century', in D. Rhind (ed.) *Framework for the World*, Cambridge: GeoInformation International, pp. 14–28.

Paul, F. (2008) '*Grand Theft Auto* sales top $500 million: First week score of one of the most lucrative entertainment events in history', *Reuters*, 7 May, <www.msnbc.msn.com/id/24502383/>.

Pedersen, P., Farrell, P. and McPhee, E. (2005) 'Paper versus pixel: Effectiveness of paper versus electronic maps to teach map reading skills in an introductory physical geography course', *Journal of Geography*, 104(5): 195–202.

Perkins, C. (2006) 'Mapping golf: Contexts, actors and networks', *The Cartographic Journal*, 43(3): 208–23.

Perkins, C. (2008) 'Cultures of map use', *The Cartographic Journal*, 45(2): 150–8.

Perkins, C. (2009) 'Performative and embodied mapping', in R. Kitchin and N. Thrift (eds) *International Encyclopedia of Human Geography*, London: Elsevier.

Perkins, C. and Parry, R.B. (1996) *Mapping the UK*, London: Bowker Saur.

Perkins, C. and Dodge, M. (2008) 'The potential of user-generated cartography: A case study of the Openstreetmap project and Mapchester mapping party', *Northwest Geography*, 8(1): 19–32.

Pickles, J. (2004) *A History of Spaces: Mapping Cartographic Reason and the Over-Coded World*, London: Routledge.

Rambusch, J. (2008) *Situated Play*, Linkoping: Linkoping Institute of Technology. Rød, J.K., Ormeling, F. and van Elzakker, C. (2001) 'An agenda for democratising cartographic visualisation', *Norsk Geografisk Tidsskrift*, 55(1): 38–41.

Salen, K. and Zimmerman, E. (2003) *Rules of Play: Game Design Fundamentals*, Cambridge, MA: MIT Press.

Schienke, E.W. (2003) 'Who's mapping the mappers? Ethnographic research in the production of digital cartography', in *Transforming Spaces, The Topological Turn in Technology Studies*, Darmstadt, Germany, <www.ifs.tu-darmstadt.de/fileadmin/gradkoll//Publikationen/space-folder/pdf/Schienke.pdf>.

Solnit, A.J. (1993) *The Many Meanings of Play: A Psychoanalytic Perspective*, New Haven, CT: Yale University Press.

Sparke, M. (1998) 'A map that roared and an original atlas: Canada, cartography, and the narration of nation', *Annals of the Association of American Geographers*, 88(3): 463–95.

SteveC (2008) 'Copyright Easter eggs', OpenGeoData Blog, <www.opengeodata.org/?p=287>.

Sutton-Smith, B. (1997) *The Ambiguity of Play*, Cambridge, MA: Harvard University Press.

Thrift, N. (2003) 'Closer to the machine? Intelligent environments, new forms of possession and the rise of the supertoy', *Cultural Geographies*, 10: 389–407.

Winnicott, D.W. (1971) *Playing and Reality*, New York: Basic Books.

Wood, D. (1987) 'Pleasure in the idea: The atlas as narrative form', *Cartographica*, 24(1): 24–45.

Wood, D. (1992) *The Power of Maps*, London: Routledge.

11 Ce n'est pas le monde (This is not the world)

John Krygier and Denis Wood

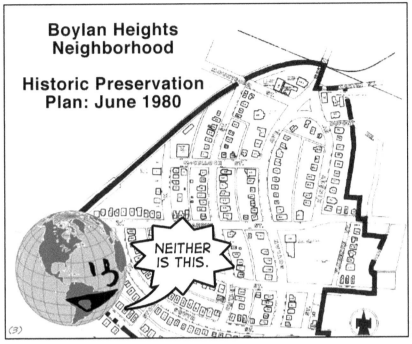

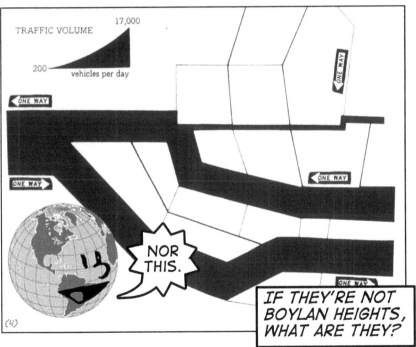

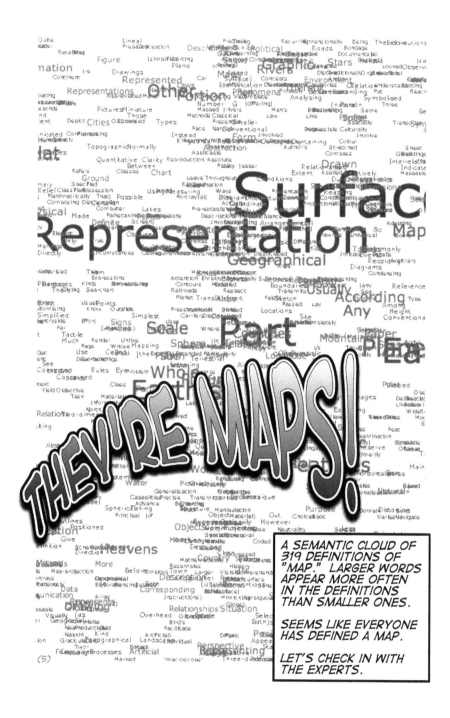

A SEMANTIC CLOUD OF 319 DEFINITIONS OF "MAP." LARGER WORDS APPEAR MORE OFTEN IN THE DEFINITIONS THAN SMALLER ONES.

SEEMS LIKE EVERYONE HAS DEFINED A MAP.

LET'S CHECK IN WITH THE EXPERTS.

(5)

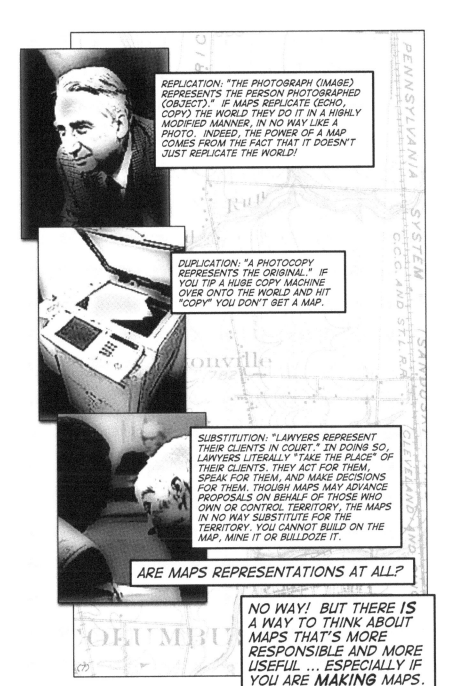

REPLICATION: "THE PHOTOGRAPH (IMAGE) REPRESENTS THE PERSON PHOTOGRAPHED (OBJECT)." IF MAPS REPLICATE (ECHO, COPY) THE WORLD THEY DO IT IN A HIGHLY MODIFIED MANNER, IN NO WAY LIKE A PHOTO. INDEED, THE POWER OF A MAP COMES FROM THE FACT THAT IT DOESN'T JUST REPLICATE THE WORLD!

DUPLICATION: "A PHOTOCOPY REPRESENTS THE ORIGINAL." IF YOU TIP A HUGE COPY MACHINE OVER ONTO THE WORLD AND HIT "COPY" YOU DON'T GET A MAP.

SUBSTITUTION: "LAWYERS REPRESENT THEIR CLIENTS IN COURT." IN DOING SO, LAWYERS LITERALLY "TAKE THE PLACE" OF THEIR CLIENTS. THEY ACT FOR THEM, SPEAK FOR THEM, AND MAKE DECISIONS FOR THEM. THOUGH MAPS MAY ADVANCE PROPOSALS ON BEHALF OF THOSE WHO OWN OR CONTROL TERRITORY, THE MAPS IN NO WAY SUBSTITUTE FOR THE TERRITORY. YOU CANNOT BUILD ON THE MAP, MINE IT OR BULLDOZE IT.

ARE MAPS REPRESENTATIONS AT ALL?

NO WAY! BUT THERE IS A WAY TO THINK ABOUT MAPS THAT'S MORE RESPONSIBLE AND MORE USEFUL ... ESPECIALLY IF YOU ARE MAKING MAPS.

and thinking through the problem in this way makes it highly questionable whether there is any sense in which maps are representations.

Maps are propositions. Maps are propositions in graphic form. Technically a proposition is a statement in which the subject is affirmed or denied by its predicate, but this is precisely what a map does, affirms the existence and location of its subjects. *This is there,* the map affirms, again and again and again.

The map's affirmation is almost uniquely robust. It says, of let us say a house of worship, that, first of all, a house of worship is a thing that exists. We call this the precedent existential proposition: *this* (a house of worship) *is.* Many maps advance existential propositions in their legends. *"House of worship,"* such legends say, and next to the words they place the mark ✝.

Maps affirm more, however, than the existence of a conceptual category. Here, maps say, is an *instance* of such a thing, and here is another, and here are others still. What makes the map's affirmation so ringing is the implicit challenge: "You don't believe it? Go check it out!" This instantiation of the precedent *existential* proposition is the fundamental *cartographic* proposition: *this is there.* We call the fundamental cartographic proposition a posting.

The posting has a logical form which is essential to observe. Just as the posting affirms the existence of the *this,* so it affirms the existence of the *there.* The *there* is the instantiation of the precedent existential proposition, *there is.* That is, the posting conjoins two precedent existential propositions, *this is* and *there is,* to give us *this is there* and, equivalently, *there is this.* Once posted, the *this* takes on *thereness,* a quality of being some*where,* as the *there* takes on *thisness,* a quality of being some*thing.*

The map's implicit challenge is not simply to find an instance of a *this* or a *there,* but to find that *this is there* precisely where *there is this.*

Most maps consist of hundreds, even of thousands of postings. Small scale maps may consist of hundreds of thousands of postings. Postings are organized by the map's propositional logic which dictates how postings may be manipulated in the construction of the map's higher order propositions. As atoms are combined into molecules, so postings are combined into higher order postings; as molecules are decomposed into their constituent atoms, higher order postings are likewise. At the atomic level, a map may post a house of worship, a school, and a street, *and* at the molecular level post a neighborhood and a city (that is, affirm the propositions that *neighborhood and city are* and *are* there). Similarly a map posting a city and a neighborhood may post a street, a school, and a house of worship. This relationship permits the conveyance of authority from one level to another.

Maps may post things that don't yet exist (as of proposed roads), or things that have ceased existing (as of ancient Rome), or that are outside the realm of existence (as of Middle-earth). One map may post a thing here that another posts somewhere else. Frequently maps propose the existence of things to which is assent is given (the Great American [...] victory in Iraq), but from which it is su[...] which new propositions are advanced (the [...] the things posted on maps – at both ato[...] instantiations of conceptual categorie[...] on the social assent vouchsafed t[...] their location.

It follows that precisio[...]

(9)

ZZZZZZZZZ...

KRYGIER FOUND WOOD'S ESSAY ENGROSSING, BUT FELT THERE HAD TO BE AN EASIER WAY TO SAY IT...

(10)

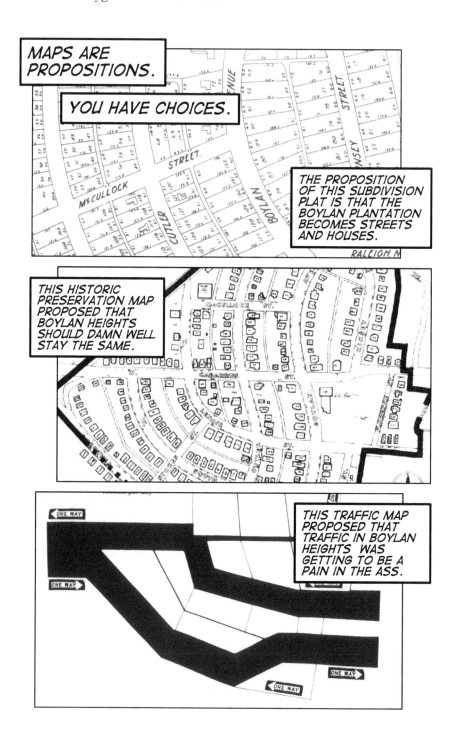

THIS JACK O'LANTERN
MAP PROPOSED THAT IN
BOYLAN HEIGHTS PEOPLE
WITH MEANS CARVED
PUMPKINS, WHILE THOSE
WITHOUT MEANS DIDN'T.

(11)

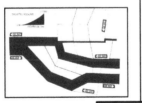

EACH MAP PROPOSITIONS
YOU WITH A DIFFERENT
BOYLAN HEIGHTS. MANY
MAPS OF THE SAME
PLACE ... YET NONE IS
BOYLAN HEIGHTS.

(12)

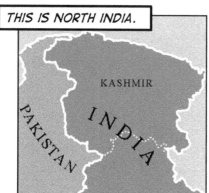

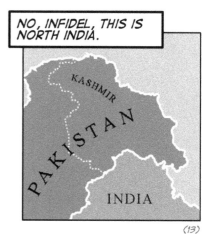

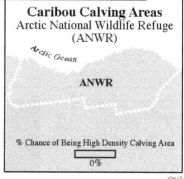

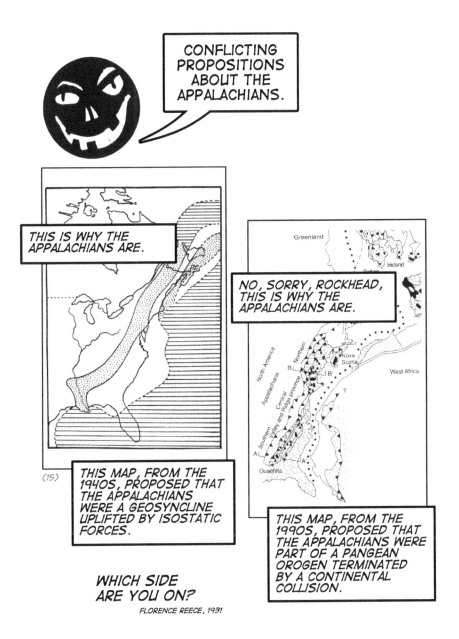

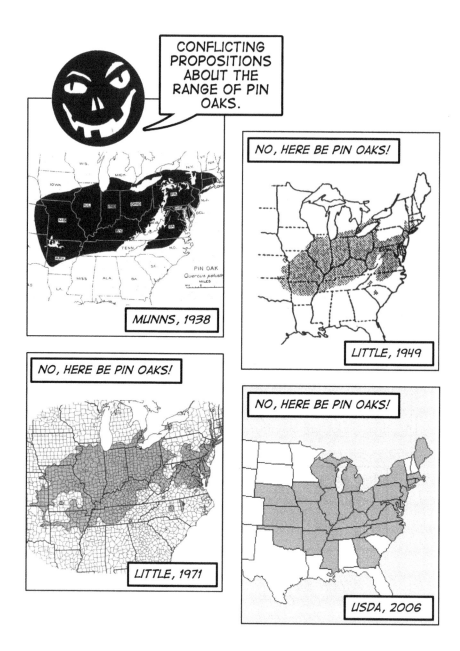

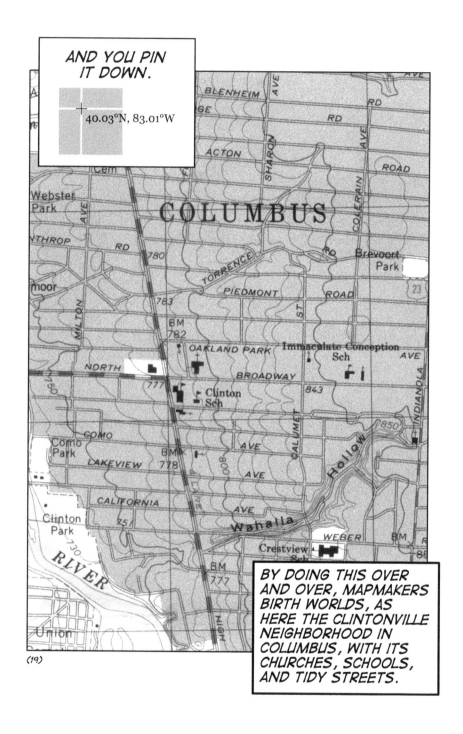

AND YOU PIN IT DOWN.

40.03°N, 83.01°W

BY DOING THIS OVER AND OVER, MAPMAKERS BIRTH WORLDS, AS HERE THE CLINTONVILLE NEIGHBORHOOD IN COLUMBUS, WITH ITS CHURCHES, SCHOOLS, AND TIDY STREETS.

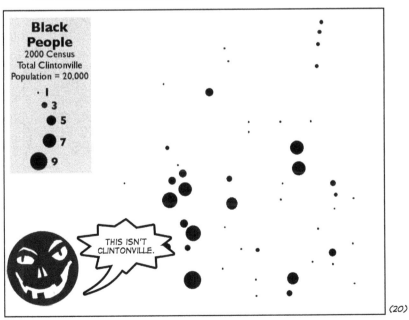

JUST BECAUSE THE REAL
WORLD CAN'T BE CAUGHT
BY A MAP DOESN'T MEAN
WE CAN'T MAP THE REAL
WORLD... AS MANY WAYS
AS WE WANT TO.

Notes

1 "Ce n'est pas le monde" is an experiment in rethinking maps and discourse about maps: a proposition about maps as propositions and about comic books as academic discourse in the form of a comic book of propositional maps. We created "Ce n'est pas le monde" in June 2006, with *Comic Life* software, presenting it at the Critical Geography Mini Conference (Columbus, Ohio) and the North American Cartographic Information Society (Madison, Wisconsin), both in October 2006, and the Geography and Humanities Symposium (Charlottesville, Virginia) in June 2007. Comments received helped us bring it to its current form, which we hope recalls the alternative comics that emerged in the 1960s (cf. Hatfield 2005) while at the same time profiting from Scott McCloud's (1993) comic-book reading of comics through the lens of C.S. Peirce's semiotics. In particular, McCloud exploits Peirce's understanding of icons, indices, and symbols (cf. Manning 1998).

2 We are far from the first to argue that maps are not representations but propositions (for example, see Acton 1938), or to allude to René Magritte's *Treachery of Images* (1928–9), his famed painting of a pipe inscribed, "Ceci n'est pas une pipe." Our map here, "Boylan Heights, Raleigh, N.C." (1908), by Boston landscape architects Kelsey and Guild, *literally* proposed Boylan Heights as a place, since before the building of the houses and the moving in of the residents, this map was the *sole* form in which Boylan Heights existed. As built, the neighbourhood realized this proposal. (Source: reproduced from the *Book of Maps 1885* (p. 114), Wake County Registry, North Carolina.)

3 "Boylan Heights Neighborhood Historic Preservation Plan" (1980) advances an alternative proposition, that of Boylan Heights as historic exemplar, "a classic early 20th century neighborhood," worthy of having its character preserved.

4 The propositions advanced by "Boylan Heights Traffic Volume" (1981), from Denis Wood's unpublished *Dancing and Singing: A Narrative Atlas of Boylan Heights*, are that traffic flowed through Boylan Heights in the volumes indicated. The *argument* advanced was that the traffic played a profound role in the neighborhood's life. The study of arguments was first given rigorous treatment in Aristotle's *Organon*. That Aristotle's syllogistic logic presupposed the more fundamental logic of propositions was established in the wake of Leibniz's work on the logical calculus, subsequently the calculus of propositions. We are attracted to the calculus of propositions because, as Bertrand Russell (1938) put it, "[a] proposition, we may say, is anything that is true or that is false," and "[t]he propositional calculus is characterized by the fact that all its propositions have as hypothesis and as consequent the assertion of a material implication." Certainly this is true of maps as well (see also Pospesel 1998).

5 J.H. Andrews (1996) collected 321 definitions of "map," dating from 1649 to 1996, in preparation for his article "What was a map? The lexicographers reply." We loaded them into the Analys.icio.us semantic cloud generator that produced the display of which we present a detail. The visual is often more effective than the verbal, a claim we make about both maps and comics (cf. Ginman and von Ungern-Sternberga 2003).

6 Here Andrews turned to three twentieth-century voices for definitions of the map. Max Eckert (1921) wrote the influential *Die Kartenwissenschaft Die Kartenwissenschaf*, but Andrews quoted from his more accessible paper, "On the nature of maps and map logic" (Eckert 1908), with the quoted remark on p. 345. Arthur Robinson was Eckert's principle intellectual heir, dominating

cartography in the second half of the twentieth century as Eckert had the first. Andrews pulled the "map is a representation of the milieu" definition from Robinson and Barbara Bartz Petchenik's (1976) *The Nature of Maps*, pp. 15–16. Andrews quoted Wood (1991) from his paper, "How maps work," the quotation on p. 66.

7 The sixth chapter of Pauline Rosenau's (1992) *Post-Modernism and the Social Sciences: Insights, Inroads and Intrusions* presents her understanding of the postmodern attack on representation that she fears makes modern social science impossible. A defender of traditional notions of representation, Rosenau presents her arguments as a "balanced appraisal." That her argument is fundamentally reactionary makes it the more pertinent for our purposes here, a succinct and encompassing survey of what people mean by representation. Our reactions to her suggestions draw on a range of sources including Richard Rorty's (1979) *Philosophy and the Mirror of Nature*; Nelson Goodman's (1978) *Ways of Worldmaking*; Michel Foucault's (1972) *The Archaeology of Knowledge*; and Paul Feyerabend's (1975) *Against Method*. We are also indebted to Andrew Pickering's (1995) *The Mangle of Practice: Time, Agency, and Science* and J.B. Harley's (2001) *The New Nature of Maps*.

8 Can an image of seduction suggest the seductive qualities of a map? Is a map *that* kind of proposition? Is it seductive? Seducing? Does the allure lie in the proposition? Or in the delusional desire for direct representation? Or both? Like "representation," "image" too implies some sort of correspondence to and mirroring of "reality" but refers to the visual more broadly. David Freedberg's (1989) *The Power of Images*, for example, deals with fine art, masks, photographs, illustrations, icons, sculpture, statuary and so on. James Elkins' (2001) *The Domain of Images* deals with fine art, pictographs, monograms, photographs, graphs, charts, indigenous paintings, schemata, money, seals, stamps, engineering drawings . . . and so on.

9 The text on these two pages abstracts a collage of the following texts: Wood's "Thinking about maps as talk instead of pictures," presented at the annual meeting of the National Council on Geographic Education in Philadelphia in 2002; Wood's "thinking about maps as propositions instead of pictures," presented at the annual meeting of the North American Cartographic Information Society in Jacksonville, Florida, in 2003; and the text, "Are maps TALK instead of pictures?" that Wood wrote for his, Ward Kaiser's and Bob Abramms' (2006) *Seeing Through Maps*. All draw on the work Wood had been doing since 2000 with John Fels on the propositional logic of the map, crystallized in Wood and Fels (2008), *The Natures of Maps*. We all – Wood, Fels and Krygier – recognize that the propositional logic of the map must be graphic, and Wood and Fels develop their "spatial/meaning calculus" graphically.

10 Visual rhetoric and comics studies seriously engage the visual in a manner appropriate to our thinking about maps as propositions (cf. Handa 2004). Visual rhetoricians ask questions such as how and why we argue visually, how we understand the myriad visual arguments aimed at us, and how we ourselves can become better at visual arguments (also see Lunsford and Ruszkiewicz 2006, especially their chapter on visual arguments). Visual rhetoric also makes strong links to semiotics and related approaches to understanding and interpreting diverse visual materials, art, advertising, movies, comic books, photographs, graphs, house plans and maps (cf. Hill and Helmers 2004). The idea of visual expressions as arguments, indeed as propositions, runs throughout the visual rhetoric literature. Comics studies are newer, but already an interdisciplinary field with

conferences, journals (*The Comics Journal, International Journal of Comic Art, Image & Narrative,* and *ImageTexT,* the latter two online), academic centres (for example, at Michigan State, Ohio State and Bowling Green State Universities), and reflective texts (in an already enormous literature numerous: Barker 1989; Carrier 2000; Gordon 1998; Heer and Worcester 2004; Inge 1990; Lefevre and Dierick 1998; Pustz 1999; Varnum and Gibbons 2001; Wright 2001. Academic acceptance of comics studies was in large part spurred on (and exemplified) by Scott McCloud's (1993) *Understanding Comics,* a comic book about comics as academically sound as it is approachable. McCloud's work in comic-book form (including his subsequent *Reinventing Comics* (2000) and *Making Comics* (2006)), established the fact that the comic form can work as intellectual discourse, a *visual* intellectual discourse. A case for geographers engaging comic books, at least at an interpretive level, has been made by Jason Dittmer (2005, 2007) who situates research on comic books as part of a broader interest in the visual components of popular culture. Our proposition about the comic as an appropriate form of academic discourse (like textual articles or verbal presentations) raises many questions. What is *wrong* with the visual that makes it so inappropriate as formal academic discourse? Why do scholars who study the visual (maps) express themselves primarily with text/words? Could a comic, a map or any other largely non-textual expression be considered appropriate as academic discourse (without the need to use notes, like these, to explain everything with words?)

11 Like the traffic map, the pumpkin map is from Wood's unpublished *Dancing and Singing: A Narrative Atlas of Boylan Heights,* although this particular map *has* been previously published, where the argument it advances is made explicit in a comparison with a map of some of the contents of the Boylan Heights neighbourhood newsletter (see also Harmon 2004: 104–7). You can also *hear* Wood make the argument in a radio interview with Ira Glass on Glass's *This American Life* (archived at www.thislife.org, selected maps from the Boylan Heights atlas can be found on the Making Maps blog: http://makingmaps. net).

12 Each map proposes a different Boylan Heights, which is precisely why the atlas Wood has been working on will contain over a hundred maps of the neighbourhood, though it would take thousands more to really begin to close in on something that, in the end, can never be caught.

13 Each map proposes a different "Kashmir." Maps of disputed territories are very easy to accept as propositional maps because "everyone" acknowledges that boundaries and territory are human constructs.

14 Each map proposes a different region of caribou calving in the Arctic National Wildlife Refuge. A rogue US Government employee placed maps of alleged caribou calving areas in the Refuge on his website. The official US Government position was that caribou calving in the Refuge is not well understood, and should not be publicized. It also denied that the employee who created the maps was an expert on caribou calving. There is thus no "official" US map of caribou calving in the Refuge.

15 One of the reasons it's important to show these maps is because "hard science" is so often the redoubt of choice for those defending the representational character of maps. "Sure," they say, "that may be true of national boundaries or in political squabbles, but those aren't scientific maps." But "scientific" maps are not a whit less propositional. When Wood took geology in college isostatic rebound was the argument advanced for the uplift of mountains. A generation later this notion

makes people smile indulgently, as at the foibles of a toddler. If you line maps
up chronologically, you see continuous change in the way humans think about
things. It's plain to us how ... *wrong* our ancestors were, and how completely
speculative, hypothetical, *propositional* their thinking was. Why do we imagine
we're any different, imagine that we finally know how the world *really* is, when
all those before us have been so misguided? The map of the geosyncline is from
O.D. von Engeln's (1949: 341) *Geomorphology: Systematic and Regional*, where
it is added in the caption that: "[t]he inferred extension seaward of the ancient
land mass, Appalachia, is indicated by a dashed line." The map of the Pangean
orogen is from Eldridge Moores and Robert Twiss, *Tectonics* (1995: 357), where
the caption calls it a "[m]ap of Appalachian-Caledonian-West African mountain
system."

16 Here eight propositions, from a multitude, about the range of the pin oak.
    See the exhaustive discussion of these range maps in Wood and Fels (2008:
    146–63).

17 Here we enter the realm of the sign since, after all, the posting is constituted
    of a sign on the cartographic sign plane. We follow de Saussure, Barthes
    and Eco, among others, in taking a sign to be compounded of a signifier and a
    signified. This is Barthes's (1973: 39 and 41) definition of the linguistic sign
    and his definition of the semiological sign-function. Umberto Eco (1976: 48
    and 49) defines a sign as "an element of an *expression plane* conventionally
    correlated to one (or several) elements of a *content plane*," though he insists,
    "[p]roperly speaking there are not signs, but only sign-functions ... realized
    when two *functives* (expression and content) enter into a mutual correlation."
    Both derive their definitions directly from Ferdinand de Saussure's (1959: 67)
    Course in General Linguistics where he says, "I call the combination of a
    concept and a sound-image a *sign*," and later, "I propose to retain the word *sign*
    to designate the whole and to replace *concept* and *sound-image* respectively by
    signified and *signifier*." The signified (concept, content, categorical type or
    whatever we're going to call it), resides in some sort of conceptual space,
    conceptual universe, content space, content plane or semantic field. This is what
    we're attempting to suggest here in this ... *evocation* ... of a semantic cloud.
    In order to evoke it, of course, we've had to marry the concepts (house of worship,
    worship, house) to ... *marks*, and so in fact these are signs (in fact, actually,
    the pertinent expressive elements of signs), *not* concepts. We know this, but let's
    pretend. (For an interesting, and occasionally hilarious, account of Saussure's
    fate at the hands of Chomsky, Barthes, Derrida *et al.*, see Roy Harris's (2001)
    Saussure and His Interpreters, not that we buy into all of Harris's complaints
    either.)

18 Here we've attempted to evoke the plane of expression, the graphic potential,
    the field of marks, the domain of signifiers, or of visual-images (in describing
    the signifier as a "sound-image" Saussure revealed his focus on speech and
    language), more successfully we feel, since this realm is material to begin
    with.

19 Presumably these signs lead to some kind of action. Why else make signs,
    why else advance propositions, unless to affect the behaviour or state of
    another? Without this motivation it is hard to understand why people would
    make, publish and disseminate maps. The sign theorist who made this point
    most straightforwardly was Colin Cherry (1957: 306) who defined a sign as "a
    transmission, or construct, by which one organism affects the behavior or state
    of another, in a communication situation." Contrast his definition of a sign with

those of Saussure, Barthes and Eco that we just gave. It's as though they came from different worlds, which in a way they did. Although written in the 1950s his text is wholly Peircean in spirit, and indeed his definition of a sign is a generalization of Peirce's, which Cherry (1957: 220) distinguishes, "by the requirement that a sign must be capable of evoking responses which themselves must be capable of acting as signs for the same (object) designatum." Peirce's sign formed an essential part of his idea of logic – his approach was philosophical not linguistic like de Saussure's – and a sign, he said, was "something which stands to somebody for something in some respect or capacity" (from Hartshorne and Weiss (1931: 228), but see also pages 227, 231, 303 and 418). Peirce distinguished three triadic semiotic relations of significance, of which the second trichotomy of the sign consisted of his famous icon, index and symbol (which we referred to earlier), *ad infinitum* (almost literally, since Peirce identifies sixty-six classes of signs). No matter how deep we dive we won't be finding many points of contact between Peircean and Saussurian signs, nor between Cherry's somewhat individual transformation of the Peircean sign and the Saussurian sign. (In fact the only connection between Cherry and de Saussure that we can point to is their joint appearance in a few paragraphs in Roman Jakobson's (1961) *Linguistics and Communication Theory*, where he is explicitly attempting a wedding, in fact a shotgun wedding, which didn't take. It's too bad, because the Saussurian sign lacks the motivation of Cherry's sign, and Cherry's sign desperately needs de Saussure's clarifying and simplifying formalism. Meanwhile, analytic linguistic speech-act theorists such as J.L. Austin (1962) are in a third world altogether, which again is too bad, because Austin's efforts at understanding *what one is doing in saying something* – especially his concept of the performative (yes, it originated here) – would be so much more valuable if they ever made contact with communication theory and/or semiology. Understanding how maps work – and how they accomplish work – really requires Peirce's and Cherry's motivation, de Saussure's sign and Austin's performativity.

20 The differences in motivation behind these two additional propositions about Clintonville reflect a resident's critical perspective of the gentrifying, stereo-typically progressive 1920s neighbourhood in the city of Columbus. The map of political contributions reveals that a *few* Republican donations are as large as *many* Democratic donations. This suggests the need for debate about campaign financing based on the imbalance of wealth. More to the point, the map proposes that, within one of the most progressive neighbourhoods in Columbus, there are a handful of wealthy Republicans who may be held partially responsible, from a local perspective, for the diverse failures of the Bush administration. Stop by and ask them how they justify their financing of the administration. The map of black residents of Clintonville proposes that the progressive residents think about the fact that racial diversity in the neighbourhood is low. Clintonville is typical of progressive, gentrifying neighbourhoods, where the politics are loud but the "practical" worries about property values and schools – code words intimately tied to race – trump politics. Both maps actively propose action – as with the topographic map that precedes them, engaging neighbours about the effects of their political contributions and addressing the contradiction of politics and diversity.

References

Acton, H.B. (1938) 'Man-made truth', *Mind*, NS 47(186): 145–58.

Andrews, J.H. (1996) 'What was a map? The lexicographers reply', *Cartographica*, 33(4): 1–11.

Austin, J.L. (1962) *How To Do Things With Words*, Oxford: Oxford University Press.

Barker, M. (1989) *Comics: Ideology, Power, and the Critics*, New York: Manchester University Press.

Barthes, R. (1973) *Elements of Semiology*, New York: Hill and Wang.

Carrier, D. (2000) *The Aesthetics of Comics*, University Park, PA: Pennsylvania State University Press.

Cherry C. (1957) *On Human Communication*, Cambridge, MA: MIT Press.

de Saussure, F. (1959) *Course in General Linguistics*, New York: Philosophical Library.

Dittmer, J. (2005) 'Captain America's empire: Reflections on identity, popular culture, and post-9/11 geopolitics', *Annals of the Association of American Geographers*, 95(3): 626–43.

Dittmer, J. (2007) 'The tyranny of the serial: Popular geopolitics, the nation and comic book discourse', *Antipode*, 39(2): 247–68.

Eckert, M. (1908) 'On the nature of maps and map logic', *Bulletin of the American Geographical Society*, 40(6): 344–51.

Eckert, M. (1921) *Die Kartenwissenschaft Die Kartenwissenschaf: Forschungen und Grundlagen zu einer Kartographie als Wissenschaft*, Berlin: W. De Gruyter.

Eco, U. (1976) *A Theory of Semiotics*, Bloomington, IN: Indiana University Press.

Elkins, J. (2001) *The Domain of Images*, Ithaca, NY: Cornell University Press.

Feyerabend, P. (1975) *Against Method*, London: New Left Books.

Foucault, M. (1972) *The Archaeology of Knowledge*, New York: Pantheon.

Freedberg, D. (1989) *The Power of Images: Studies in the History and Theory of Response*, Chicago, IL: University of Chicago Press.

Ginman, M. and von Ungern-Sternberga, S. (2003) 'Cartoons as information', *Journal of Information Science*, 29(1): 69–77.

Goodman, N. (1978) *Ways of Worldmaking*, Indianapolis, IN: Hackett.

Gordon, I. (1998) *Comic Strips and Consumer Culture, 1890–1945*, Washington DC: Smithsonian Institution Press.

Handa, C. (2004) *Visual Rhetoric in a Digital World*, Boston, MA: Bedford/ St. Martins.

Harley, J.B. (2001) *The New Nature of Maps*, Baltimore, MD: Johns Hopkins University Press.

Harmon, K. (2004) *You Are Here: Personal Geographies and Other Maps of the Imagination*, New York: Princeton Architectural Press.

Harris, R. (2001) *Saussure and His Interpreters*, New York: New York University Press.

Hartshorne, C. and Weiss, P. (1931) *Collected Papers of Charles S. Peirce*, Cambridge, MA: Harvard University Press.

Hatfield, C. (2005) *Alternative Comics: An Emerging Literature*, Jackson, MI: University Press of Mississippi.

Heer, J. and Worcester, K. (2004) *Arguing Comics: Literary Masters on a Popular Medium*, Jackson, MI: University Press of Mississippi.

Hill, C. and Helmers, M. (2004) *Defining Visual Rhetorics*, Mahwah, NJ: Lawrence Erlbaum.

Inge, M. (1990) *Comics as Culture*, Jackson, MI: University Press of Mississippi.

Jakobson, R. (1961) 'Linguistics and communication theory', in Jakobson, R. (ed.) *On the Structure of Language and Its Mathematical Aspects*, Providence, RI: American Mathematical Society, 245–52.

Lefevre, P. and Dierick, C. (1998) *Forging a New Medium: The Comic Strip in the Nineteenth Century*, Brussels: VUB University Press.

Lunsford, A. and Ruszkiewicz, J. (2006) *Everything Is an Argument*, 4th edn, Boston, MA: Bedford/St Martins.

McCloud, S. (1993) *Understanding Comics*, Northampton, MA: Kitchen Sink Press.

McCloud, S. (2000) *Reinventing Comics*, New York: Paradox Press.

McCloud, S. (2006) *Making Comics*, New York: Harper.

Manning, A. (1998) 'Scott McCloud – understanding comics', *IEEE Transactions on Professional Communication*, 41: 60–9.

Moores, E. and Twiss, R. (1995) *Tectonics*, New York: Freeman.

Pickering, A. (1995) *The Mangle of Practice: Time, Agency, and Science*, Chicago, IL: University of Chicago Press.

Pospesel, H. (1998) *Introduction to Logic: Propositional Logic*, Upper Saddle River, NJ: Prentice Hall.

Pustz, M. (1999) *Comic Book Culture: Fanboys and True Believers*, Jackson, MI: University Press of Mississippi.

Robinson, A.H. and Petchenik, B.B. (1976) *The Nature of Maps: Essays Toward Understanding Maps and Mapping*, Chicago, IL: University of Chicago Press.

Rorty, R. (1979) *Philosophy and the Mirror of Nature*, Princeton, NJ: Princeton University Press.

Rosenau, P. (1992) *Post-Modernism and the Social Sciences: Insights, Inroads, and Intrusions*, Princeton, NJ: Princeton University Press.

Russell, B. (1938) *Principles of Mathematics*, 2nd edn, New York: Norton.

Varnum, R. and Gibbons, C. (2001) *The Language of Comics: Word and Image*, Jackson, MI: University Press of Mississippi.

von Engeln O.D. (1949) *Geomorphology: Systematic and Regional*, New York: Macmillan.

Wood, D. (1991) 'How maps work', *Cartographica*, 29(3/4): 66–74.

Wood, D. and Fels, J. (2008) *The Natures of Maps*, Chicago, IL: University of Chicago Press.

Wood, D., Kaiser, W. and Abramms, B. (2006) *Seeing Through Maps: Many Ways to See the World*, Amherst, MA: ODT.

Wright, B. (2001) *Comic Book Nation: The Transformation of Youth Culture in America*, Baltimore, MD: Johns Hopkins University Press.

Ce n'est pas une citrouille avec une pipe.

12 Mapping modes, methods and moments

A manifesto for map studies

Martin Dodge, Chris Perkins and Rob Kitchin

Introduction

By way of conclusion to *Rethinking Maps* we want to set out a manifesto for map studies for the coming decade. Its goal is to generate ideas and enthusiasm for scholarship that advance our understanding of the philosophical underpinnings of maps, and also enhances the practices of mapping. This is not a call for ever more introspective intellectual navel gazing about maps. Instead it traces routes and methods that might help people to *do* mapping differently and more *productively*, in ways that might be more efficient, democratic, sustainable, ethical or even more fun. This manifesto is, of course, preliminary and partial, coming as it does from a social scientific tradition and the authors' experiences as Anglophone human geographers. It also focuses on understanding *everyday* mapping practices and the various *socio-technological infrastructures* that are a necessary, but often unquestioned, support for contemporary mapping. The aim is to suggest and provoke. Our manifesto for map studies is structured into three 'levels': first, looking at *modes* ('what to study'); second, *methods* ('how to study'); and finally, *moments* ('when and where to study').

Modes of mapping

For us, map studies needs to continue to develop alternative ways to think through cartographic history and contemporary practice that are not wedded to simplifying, modernist, narratives of 'advancement'. In this pursuit, we might build on the relational thinking of Matthew Edney. He forwards the notion of 'cartography without progress' (1993: 54), in which mapping is read as 'a complex amalgam of cartographic *modes* rather than a monolithic enterprise'. For Edney, a cartographic mode is not simply a linear chronological sequence, instead it is a unique set of cultural, social, economic and technical relations within which cartographers and the map production processes are situated. The mode is thus the milieu in which mapping practices occur. Each cartographic mode gives rise to its own kind of map artefacts, and critically this conceptualization does not assume that one is inherently better than another, or that one mode will inevitably evolve into a 'superior' mode. As Edney (1993: 58) elaborates: '[t]he mode is thus the combination of cartographic form and cartographic function, of the internal construction of the data, their representation on the one hand and the external *raison d'être* of the map on the other.'

Modes are unique to their time and place, and are transitory. Modes of mapping practice are coupled to the continual emergence of new knowledges, spatial problems, methods and institutions, and drive developments in the design of map representations and in the roles that maps play in society. There are usually multiple but distinct mapping modes operating at the same time in the same place. Modes can interact and may well overlap, merge and diverge. The boundaries between them are likely to be fuzzy and permeable. Cartographic history, according to Edney's theorization, is therefore best read as a plural and relational network of activities, rather than a single linear process. In contemporary cartographic epistemologies, a diverse range of mappings is seen to emerge from a shifting creative milieu, the end result of which is not a unidirectional evolutionary tree of maps, but rather a complex, many-branching, rhizomatic structure.

Part of the undoubted excitement at the moment about maps stems from the fact that contemporary mapping practices consist of multiple, overlapping modes. Mapping is emergent and variegated, drawing on many disparate ideas and data sources, produced by a diverse collection of practitioners and activists, utilizing many forms of visualization. Mapping is thoroughly situated in wider socio-technical changes (particularly the diffusion of the Internet throughout map production and the use of the Web as the main medium of dissemination). To begin to excavate the nature of contemporary mapping modes requires empirical analysis to unpack cultural, social and technological relations that determine these cartographic practices. It seems to us that it would be productive for researchers to focus attention on: (i) interfaces, (ii) algorithms, (iii) cultures, (iv) authorships and (v) infrastructures.

Interfaces: mapping out screen spaces

More and more, everyday mapping is encountered as part of a digital interface, or the map is itself an interface that can be queried. These 'screen spaces' are becoming an important site for analysis in map studies. What are the cultural, social and economic relations that bring the interface into being? Interrogating the interfaces of mapping is an ontological project with political ramifications. There is an emerging body of work on the critical reading of computer interfaces that can be drawn upon and might inform map study. For example, Selfe and Selfe (1994: 485) argue that one can hermeneutically read what gets screened as if it is a cultural map that 'order(s) the virtual world according to a certain set of historical and social values that make up our culture'. Interfaces en-frame and exclude, working as mediating windows onto the world. The task of decoding the embedded cultural biases and distortions in processes of interface screening is challenging, even for supposedly 'open' Web mapping interfaces because, as Parks (2004: 39) notes, they 'tend to keep users naïve about the apparatus that organizes and facilitates online navigation and how its processes occur in time and extend across space'.

Beyond the cultural politics within spaces of display, there are also phenomenological considerations relating to interfaces (cf. Introna and Ilharco 2006). Mapping often dynamically updates to reflect embodied position and kinetics (Willim 2007), inviting interrogation of the differences digital interfaces make to individual identity and social behaviour that stem from 'being on the screen'. This interface between person, map and the world in motion would once have been reserved for specialized and particularly military applications, but is now the everyday experience for many when walking with a mobile phone, driving with a satnav, flying with the airshow maps on an in-flight entertainment system, and even playing with handheld GPS units in treasure hunting games of geocaching.

Algorithms of mapping

As outlined above, the technological practices of map representation are increasingly rendered through computer interfaces on digital screens. What lies beneath these interfaces? They are all products of software, continuously brought into being by a complex amalgam of data and algorithms. These codes are highly technical but also deeply culturally contingent, yet from an investigative point of view they are very hard to read or critique.

Map studies needs to open the 'black boxes' of mapping software, to start to interrogate algorithms and databases, and in particular to investigate the production of ready-made maps that appear almost magically on the interfaces of gadgets and devices we carry and use everyday, often without much overt thought about how they work and whose map they project onto their interface. This agenda was aptly expressed by Laura Kurgan (1994: 17) in her imaginative work examining the inherent indeterminacy of the inner workings of GPS software from the external mappings its produces:

> [b]ut the space or the architecture of the information system that wants to locate and fix us in space has its own complexity, its own invisible relays and delays. The difficulty of charting the spaces that chart the spaces, of mapping the scaleless networks of the very system that promises to end our disorientation, demands redefining the points and lines and planes that build the map, and lingering in their strange spaces and times.

Opening the 'black box' of cartographic algorithms was a core element of the social science critique of GIS in the 1990s (Pickles 2004). The rapid popularization of digital mapping in the last five years makes this even more pertinent for map studies, as millions of people walk and drive around with what are effectively mini-GIS mapping gadgets in their pockets and on their vehicle dashboards.

It seems there are several productive routes to critique mapping codes. First, we can draw on emerging ideas in the field of 'software studies' that treat code as a form of material culture that can be examined from multiple points of reference to reveal how it comes into being, and works often automatically and autonomously in the world. These ideas seek an expanded understanding of software beyond the technical. They also critique how the world itself is captured within code in terms of algorithmic potential and formal data descriptions (cf. Dodge and Kitchin 2009). This research is trans-disciplinary, often driven by scholars and intellectual hackers in media theory and new media art. Fuller (2008: 2) argues that this kind of approach: 'proposes that software can be seen as an object of study and an area of practice for kinds of thinking and areas of work that have not historically "owned" software, or indeed often had much of use to say about it.' There is much, we believe, that needs to be said by people who have traditionally not 'owned' mapping codes.

Socially rich work investigating the spatiality of software algorithms and data structures has begun in human geography, notably with Thrift and French's (2002) theorization of the 'automatic production of space' and Graham's (2005) discussion of the socio-geographical effects of 'software sorting'. However, analysing algorithmic processing underlying new forms of online mapping has so far received little attention. A noteworthy exception is Zook and Graham's (2007) work on 'digiplace' as the mapped interface arising from the opaque complexity of search engine databases and spatial-relevance ranking algorithms. This research offers a significant opening and needs to followed-up and expanded upon.

A second route toward analysing mapping algorithms is to build explicit connections between cartography and the emerging conceptual agenda of 'surveillance studies' to reveal the social power frozen in code and the dangers of discriminatory effects emerging from automated sorting of people and code-based representations of place. There is a focus on power at the heart of 'surveillance studies' according to Lyon (2007: 1) with explicit attempts to explain surveillance practices in terms of 'rationalization, the application of science and technology, classification and the knowledgeability of subject'. Considering the computerized map as a surveillant technology was initially undertaken by Pickles (1991) who argued that nation states, trans-national corporations and the interests of capital and technology deploy the surveillant potential of mapping to restructure local, regional, national and global geographies. Notable examples of recent work in this vein includes Crampton (2004) who explored parallels between the nineteenth-century emergence of crime mapping and contemporary post-9/11 surveillance discourses as reflections of Foucauldian rational governance.

Mapping visual culture

In the 1990s, a research focus on the analytical functions of GIS led to a significant retreat from design issues that had hitherto formed a central concern for cartographic research. It has been argued that this retreat has almost led to the death of cartography as a discipline (Wood 2003). Everyday mapping however, grew apace from the end of the decade, and meanwhile a newly energized emphasis on the visual pervades much critical thought across cultural and media studies (see Sturken and Cartwright 2007 for an overview in this area). We would argue that a new and critical engagement with visual studies could usefully inform research into mapping. Cartography may or may not be heading toward extinction as a technical discipline, but mapping is very much alive and technology alone is insufficient an explanation for the role that new kinds of mapping are playing in society.

Such research might usefully explore new ways of envisioning spatial data in interactive and animated systems, building on the innovative work carried out by researchers such as Dykes *et al.* (2005). Which new ways of symbolizing data work best? Which widgets offer the most appealing ways of performing screen navigation and selection and why? How might geovisualization best represent movement, change and dynamic data? What are the best ways of situating the observer on and in mapping displayed on different kinds of device? Among areas that might usefully receive attention here are the interplay between screen design issues and display design issues: a much greater contextual awareness of the intertextuality of displays could inform critical approaches to the burgeoning literature around usability (see van Elzakker *et al.* 2008). Although a concern with designing better maps has led to a profusion of expert systems encouraging effective use of industry-standard software designs (e.g. Harrower and Brewer's (2003) innovative work on Colorbrewer tool), innovative design solutions for the representation of phenomena only rarely feed through to the mainstream consumption. Yet the immediate appeal of Google Earth stems in large part from the visual novelty of its interface. Mapping researchers could usefully learn from this approach. The difference that media make is also a rich research area: interesting work is already exploring the roles that sound and taste mapping might play in multimedia map design (cf. Taylor 2005).

To realize this kind of research result, mapping needs to be situated in relation to other media. It is noticeable that the mainstream of visual culture and visual studies research almost completely elides mapping at present (see for example, Elkins 2003) and that mainstream visualization research largely remains grounded in scientific representation (see Dodge *et al.* 2008). Critical insights from visual studies, with its emphasis upon innovative methodologies could usefully be applied in the more narrowly defined worlds of geovisualization. Researchers might learn much here from the practical worlds of computer game design and some of the roles that maps play in these (see for example, Longan 2008 for a critical examination of mapping/landscape relations in role-playing games where maps are so much more than a neutral backdrop for the action). Surely dialogue between visual studies and cartography would yield richer and more complex insights into the nature of mapping.

Authorship of mapping

It is also important, we believe, to focus attention in map studies on authorship. Significant changes in notions of authorship are at the heart of many contemporary modes of mapping. In particular there is a fracturing of authorship with the emergence of a more 'writerly' kind of mapping (following Roland Barthes), which according to Pickles (2004: 161) can 'engage the reader as an "author" and insist upon the openness and intertextuality of the text'. Moreover, many aspects of mapmaking practices are undergoing a metamorphosis towards a 'remix' cultural model of production that is apparent in many other media (cf. Bolter and Grusin 1999; Diakopoulos *et al.* 2007), in which new media constantly reinterpret existing media in a process facilitated by rapid and unconstrained access. Manovich (2005: no pagination) argues that '[r]emixability becomes practically a built-in feature of digital networked media universe.'

Research needs to consider the implications for mapping. How do new models of map authorship work in practice, for example: How are power structures altered by the rise of the amateur mappers? How do crowds generate wisdom in cartography? To what extent is the democratization of production really taking place? How might map 'hackers' fashion genuinely useful hybrid forms of cartography as opposed to merely creative experiments with little lasting value? Who are the new collaborative authors and why are they motivated to map? and What kinds of mapping do they do and is that mapping of quality and utility to others?

The rise of map mashups has been a significant marker of changing authorship and possibly a new mode of mapping that Crampton (2008) has termed 'Maps 2.0' (cf. Geller 2007; and Gartner this volume for useful overviews). Mapping mashups are websites or Web applications combining content from more than one source to serve a new service, and usually depend upon a third party releasing an application programming interface. We might usefully investigate the pragmatic effects and wider political implications of the emergence of these new ways of mashing maps together. Are they a relatively transitory burst of creativity that will fade as most users return to a few maps produced by high-profile institutions, or do they herald the beginnings of a lasting 'prosumer'[1] revolution? The deeper motivations for being a prosumer, and the degree to which these changes will create trusted and reliable mapping are still largely unknown.

The vanguard of prosumer authorship however, lies beyond mashing together existing data. Instead it offers newly made and often collaborative geospatial data under the guise of FOSS ('free and open source software') doctrines. The authorship of so-called 'open-source' mapping has a strongly counter-cultural ethos, itself a mixing of libertarian freedom of access to information, the socially progressive benefits of non-profit production and opposition to corporate capitalism. Of course it is ironic that much of this work is currently heavily reliant on the GPS system, designed, funded and maintained by the US military. Prosumer mapping has emerged outside of mainstream cartography, driven by enthusiastic and loosely coordinated collectives of activists, artists and programmers. Most have no formal cartographic training or professional GIS credentials, just an interest in the geography in its common-sense meaning, a liking for maps, a deep affinity with technology and, above all, passion for hacking their own elegant solutions; indeed, one of the first books to formalize the field is called *Mapping Hacks* (Erle *et al.* 2005).

Open-source authorship changes who can make maps and how they are made and open-source mapping seeks to harness the tremendous productive potential of mass-participation (the so-called 'crowd-sourcing' methodology). Such 'bottom-up' volunteer knowledge creation (seen elsewhere, for example, in Wikipedia) exploits the collaborative capacity of the Web and seeks to remake mapmaking as a social activity. Open-source mapping potentially becomes a way of thinking critically about the *practices* of cartography and not the end *products*. The map is not revered and reified as a special-knowledge product (akin to the 'Master Map' as Ordnance Survey markets its main digital topographic product) created by an elite organization and then used by a select few. Instead it becomes something that can be creatively made by many hands and enjoyed by anyone and everyone, without onerous and restrictive licencing. In the particular context of British mapping infrastructure for example, this ethos is mixed with a distinctly anti-establishment streak focused on the longstanding critique of Ordnance Survey's monopolistic pricing/licencing model, which has effectively excluded many individuals, non-profit groups, small businesses and local communities (Dodson 2005). This restrictive local context has certainly been a spur to citizen cartographers aiming 'to build a set of people's maps: charted and owned by those who create them, which are as free to share as the open road' (Dodson 2005, no pagination). Open-source mapping alternatives increasingly represent a direct challenge to the closed-world of cartographic officialdom, with its unaccountable state authorship, its emphasis upon owned and protected products as capital assets, and its claims to provide an exclusive topographic text that spatially prescribes so many aspects of daily life.

Within the domain of authorship map studies might also explore so-called 'counter-mappings' (see Harris and Hazen this volume), to pin down the scope of genuinely alternative, subversive and emancipatory mapmaking and the degree to which this mapping has effect. For example, one could argue that much open-source mapping is actually not radical at all – it simply recreates a mirror copy of existing topographic mapping, albeit distributed under a more egalitarian licencing regime. Is it possible to author counter-mappings that really challenge established power relations and effect political change? Pickles (2004: 185), for example, invokes the work of William Bunge, which he typifies as a nomadic counter-cartography, with its '[s]imple maps of hazardous materials along streets, incidences of rat-bites, or unlit alleyways'. But did Bunge's map really help 'take-back' the streets by empowering communities?

Infrastructures of mapping

The fifth and final domain through which map studies can investigate contemporary modes of mapping is to engage with infrastructure. Despite the fact that '[i]nfrastructure can be dullest of all topics', Norman (1998: 55) notes '[i]t can also be the most important. Infrastructure defines the basis of society; it is the underlying foundation of the facilities, services and standards upon which everything else builds'. Critical interrogation of the infrastructures of everyday living has been widely overlooked by the social sciences because of the ways they tend to slip beneath the surface (Graham and Thrift 2007; Star 1999). Infrastructure is often materially unseen and hidden from view; most users are unaware of it and have no experience of its significance in their everyday lives; technical systems are largely ignored as banal and 'taken-for-granted'; and infrastructure is hard to analyse because complex corporate ownership structures and fragmented regimes of regulation in the wider neo-liberal political economy tend to mask its existence. From a political perspective, critical studies of infrastructures are made more difficult because of the ways in which institutions deliberately structure them as 'black-boxed' systems to keep people from easily observing (and questioning) their design and operational logic. The invisibility of the infrastructure provides an effective cloak under which market manipulation and socially iniquitous practices can be safely carried out by institutions owning and operating them without undue negative public attention.

The lack of critical studies of mapping infrastructures tends to reify biases in the ongoing production of common cartographic data (such as topographic, routing, statistical maps) and to deny alternative ways to build and operate infrastructures. However, these infrastructures have the tendency to widen social difference and inequalities across space. As Pickles (2004: 146) argues:

> [a]s the new digital mappings wash across our world, perhaps we should ask about the worlds that are being produced in the digital transition of the third industrial revolution, the conceptions of history with which they work, and the forms of socio-political life to which they contribute.

Researching mapping as an infrastructure needs to foreground the material-ity of production, render transparent usage, and denaturalize the everyday appearance of maps by highlighting corporate structures that are underlying mapping. Working through infrastructures can be approached in two ways: first, one can consider the infrastructures that make a mapping mode possible. For example, the pivotal role of military infrastructures in everyday mapping has long been appreciated in historical studies (e.g. Harley 1988). But it important to realize that the current paths of technical development in mapping are still dependent, in large part, on military infrastructures in various guises and their significance munificence of capital and other resources (cf. Cloud 2002; Kaplan 2006). In particular the underlying geospatial capture infrastructures, such as earth imaging and GPS, are strongly influenced by military funding and imperatives of state security and secrecy. A recent example reported in the press amply illustrates this, with the launch in September 2008 of a new high-resolution commercial imaging satellite, called Geoeye, which is part supported by Google (who gain exclusive commercial access), but over half of the $502 million cost was financed by the US military. Furthermore, the Geoeye system operates under licence from the US government, who ensures their continued primary access to imagery ('shutter control') and denies highest potential resolution to anyone without explicit government authorization (cf. Chen 2008).

Second, it is important to analyse the ways in which mapping modes contribute to infrastructures themselves. The mundane disciplining role of mapping infrastructures in systems of computerized governmentality continues to grow, for example, in consumer marketing and crime mapping (Crampton 2003); this needs to be actively questioned by map studies. Rather than contributing to a more democratic society, one could argue that the powerful gaze of cartographic visualization at the heart of surveillance infrastructure means mapping is active in deepening the social power of corporations and the state over the citizen, particularly after 9/11. This is evident from the prominence of mapping in the fetishization of geospatial capabilities to 'target terrorism' (Beck 2003). A critical approach is needed here (see O'Loughlin 2005) – one research possibility is to follow the money directly from military and intelligence sources towards the mapping research that they fund. Such surveillance requirements are also a driver in the development of new mapping techniques for cyberspace, particularly for visualizing online social networks (cf. Dodge 2008).

Mapping methodologies for map studies

How can contemporary mapping practices and socio-technological infrastructures of cartography be studied empirically? What are the new methodological routes in the study of map modes? Do approaches from science and technology studies (STS), Actor-Network Theory, ethno-methodology and non-progressive genealogy that are now *de rigueur* in many areas of social science work for mapping? Can they help scholars to reconstruct the real conditions under which mapping is brought into being, or offer novel insights into how a map might make a difference in the world?

It seems clear to us that there are many valid and potentially valuable routes into the study of contemporary mapping practice. Some of these have been touched upon, in varying degrees, by the contributions to this volume (e.g. Craine and Aitken's consideration of affect; Crampton's excavation of Foucauldian genealogy; or Krygier and Wood's propositional view of mapping as situated cognitive cartography). It is, we would argue, a stimulating time for mapping scholarship with many challenges and opportunities opening up: no single epistemological position now dominates interpretation. We suggest here a range of methodological routes that might be worth pursuing, focused upon (i) materiality, (ii) political economy, (iii) affect and (iv) ethnography.

Materiality of mapping

In many other areas of the social sciences there has been a marked turn towards the materiality of objects in social processes, with a concern for the tactile experience of things, the ways this facilitates action and a focus on how the physicality of their production affords particular solutions to problems (see for example, Clark *et al.* 2008). The materiality of mapping has been largely overlooked in cartographic scholarship,[2] and in particular in contemporary research on digital products and the virtualization of interaction and experience online. In practice, paper maps are still used and many times digital maps are printed out for immediate, convenient use and annotation. Meanwhile, digital map interfaces need to be interacted with in very material ways (e.g. manipulating buttons with fingers, adjusting the position of screens to make things more visible in imperfect lighting conditions and so on). Consequently, there is a need for work that moves beyond the narrow examination of the effectiveness of 'special' tactile map products (see for example, Rowell and Ungar 2003), to interrogate *everyday* material encounters with mapping in different contexts. This needs to consider how the material forms of mapping might make a difference and perhaps explore the kinds of affordance these enable, *and* disable, and the contributions of the material in everyday problem-solving with maps.

The political economy of mapping

A major methodological element of map studies should be to explore the political economy of contemporary mapping. In the late 1980s social construc-tivist research began to interrogate the power of mapping and its historical implication in capitalist modes of production (see for example, the classic studies by Harley 1989; Harvey 1989; St Martin 1995). Similarly, there were a number of studies on the use of cartography in the propaganda of nation states and others (e.g. Monmonier 1996a). However, a political-economic approach is very rarely taken in studies of contemporary mapping, despite the fact that the vast bulk of mapping, measured in terms of volume, scale and spatial coverage, is still produced and owned by government institutions and large corporations. This concentration of spatial power is likely to remain the case into the future as well, notwithstanding the current fashion and fascination with 'open' maps made with volunteer effort. So tracing the monetary and political structures underlying the production of maps used in everyday practice is worthwhile. The fact that we seem to have more 'free' access (i.e. underpinned by advertising revenue) to detailed mapping than ever before, via Internet portals masks continuing limits to availability of large-scale data that stem from official and corporate secrecy (cf. Lee and Shumakov 2003). Decisions on where capital is being invested to produce updated and new maps, data and delivery systems affects, in practical and political terms, how the world is going to be envisioned cartographically in the future, but is opaque to scrutiny. Who controls what gets mapped when you enter a mundane geographical search query on the Web, or type a postcode destination into the find menu on your satnav, or text 'locate' on your phone? Tracing out patterns of capital investment, government subsidies, licencing fees and profits that circulate continuously, but unseen, through maps can reveal the wider power structures in which everyday mapping practice is situated, many of which are several steps removed from moments of use.

Affective understandings of mapping

Research methods also need to consider mapping as practices. Two of us have argued elsewhere that new insights will emerge if mapping is studied processionally rather than representationally (cf. Kitchin and Dodge 2007). From that perspective, there is a need for research that examines contemporary map creation as a performance of space and the affective power flowing from of-the-moment map use in diverse contexts.

There is a burgeoning body of research on the affective nature of spaces in human geography that is clearly relevant to practices of mapping (see Anderson and Harrison 2006 for a useful overview of this emerging field). This kind of research might consider: the emotional capacity of maps to do work in the world; the kinds of action and affect enabled in everyday mapping activities; and the role affect might play in enacting solutions to spatial problems. Thinking affectively could also grant insights in how people map, by focusing attention on the relations between design and its deployment, which would help professional mapmakers to create a wider range of products and interfaces capable of evoking a greater variety of actions and responses beyond the often taken-for-granted neutrality of the map as problem-solving artefact.

Thinking about what affective maps are and might be like has already begun (see Aitken and Craine 2006). Experimental examples that tap into feelings have been produced, particularly by artists (e.g. recent work around beauty mapping by Christian Nold and angry maps by Elin O'Hara Slavick 2007). In epistemological terms several scholars have begun to see the exciting and innovative potential for making mapping that encompasses affective qualities of space. For example, the recent work of Mei-Po Kwan and collaborators (e.g. Kwan 2007) enacts a feminist re-imaging of GIS as an affective and emotional alternative to neutral science, and Margaret Pearce (2008) has translated the sense of place from the narrative of trapper's diaries into affective maps of their journeys in eighteenth-century Canada.

Ethnography and novel evaluation of mapping

The need to capture *how* maps emerge into the world to do their work necessitates more nuanced means of evaluation than has typically been employed in academic cartographic research to date. Studying mapping needs to progress outside controlled laboratory environments and to seek deeper ethnographic understanding of mapping in the 'wild', so to speak. Here the focus moves from measured responses to tests towards situated observations and participation in the mapping process (see Perkins 2008). Ethnographically a map is not a map because it looks like a map, rather mapping is defined by how maps are used in practice and how they perform space. Capturing everyday mapping performance and attempting to interpolate multiple and opaque meanings is challenging conceptually and time-consuming empirically. Gaining access to natural, vernacular and everyday settings to observe situated mapping activities requires creative solutions and negotiation for scholars whose experience has mainly focused on bringing people into their labs for testing. But computer anthropologists and human–computer interaction (HCI) researchers have successfully moved in this direction in their research on how people (mis)use computers (Dix *et al.* 2004). An insightful step in this direction for map studies, which draws on experiences from HCI research is demonstrated in Barry Brown and Eric Laurier's (2005) work on the use of mapping in everyday wayfinding, in which they observe real-world navigational behaviour of people travelling in their cars. Beyond academic studies per se, another constructive illustration of the ethnographic method is Stephen Gill's (2004) photography project, which is really a visual essay resembling in many ways the mundane essence of mapping (Figure 12.1).

One area that seems ripe for such an approach is the study of the cultural practices of open-source mapping. Here, ethnographic methods could be profitably used to study key activists through participant observation of mapmaking work (such as OpenStreetMap). Work is also needed to examine the organizational structures of open-source mapping projects, the incentives for participants and the mechanisms for creating trust in the wiki production of cartographic knowledge. These could be studied as actor-networks, drawing partly on data contained in online discussion lists and blogs, to reveal the complex and contested ways that new mappings are brought into the world.

Figure 12.1 Street photography captures the immediate and embodied use of mapping for orientation and navigation. Gill's images of maps in action also reveal that often mapping is a collaborative process that involves negotiation over the map and the relation to current position and destination. Source: Ronson 2004.

It should also be possible directly to analyse the authorship of the map, because map data itself can tell stories of its own manufacture (see Figure 12.2). This effort at mapping the mappers begins to lift the lid on the traditionally anonymous authorship and authority (see above). Interestingly, this kind of analysis of authorship has already begun to reveal a lack of broad democratic participation in some open-source mapping projects (cf. Haklay 2008).

In addition, there needs to be more ethnomethodology in map studies. Such studies would focus on the use and practices of digital mapping systems and tools (e.g. satnav maps), and would research how technologies are used by different people, instead of how the systems have been designed to work. Studies would be small-scale and focused rather than generalist in nature. This kind of research could usefully study incomplete and failed mapping practices (e.g. getting beyond 'scare stories' of satnav 'blunders'; see Figure 12.3), and conflicted activities to reveal social contexts and the embodied experience of cartographic problem solving. A pragmatic end-goal of such local field studies is to reconstruct the conditions under which mapping is deployed, so as to help in the design of future map systems.

Besides ethnographic studies out in the field, we suggest that future map studies should move beyond conventional evaluative methods for revealing

Figure 12.2 The work of multiple map authors contributing to the OpenStreetMap project. Source: author-generated using ITO!'s OSM Mapper service, www.itoworld.com/static/osmmapper.

the effectiveness of cartographic representations (typically through psychological and cognitive testing in rather artificial lab settings), to look at how people manipulate and play with maps (see Perkins this volume; van Elzakker *et al.* 2008). Online three-dimensional virtual worlds and multiplayer games might become useful experimental and experiential spaces for such map evaluation. Processes of testing can be made more engaging and perhaps fun, but with the capacity for comprehensive and rigorous data capture of how users do what they do. Some steps in this direction have been taken by Michael Batty's team at the Centre for Advanced Spatial Analysis in their evaluation of thematic maps, geometric building models and spatial simulations inside virtual worlds (Batty and Hudson-Smith 2007).

The moments of mapping

In this third section of a manifesto for map studies we want to think through when and where mapping really matters. How can scholars identify some of the significant times and places of mapping practice that need to be examined in detail? Instead of the usual and sometimes sterile enumeration of particular sectors, contexts, cultures, places or even types of map or product, we argue that a focus on key processes is more likely to reveal critical aspects of mapping. As such, we offer a tentative list of mapping moments that we think are significant and worthy of study: (i) places and times of failures, (ii) points of change, (iii) time–space rhythms of map performance, (iv) the memories of mapping, (v) academic praxis; and (vi) newly creative engagement with mapping practice.

Moments of mapping failure

The moment when things go wrong often highlights how things really work, a point often overlooked in everyday life. For example, how a software glitch in an air traffic control system leads to the grounding or re-routing of all planes flying in that sector (Dodge and Kitchin 2004). These moments of failure are revealing of the world in process. As Graham and Thrift (2007) discuss, infrastructures – and as noted above mapping is in many respects an informational infrastructure of contemporary capitalism – are often most easily exposed to critical scrutiny when they fail; '[p]erhaps we should have been looking at breakdown and failure as no longer atypical and therefore only worth addressing if they result in catastrophe and, instead, at breakdown and failure as the means by which societies learn and learn to re-produce' (Graham and Thrift 2007: 5).

Many breakdowns in utility and reliability of digital mapping can be related to errors in software code that brings the map to the screen. Often these breakdowns are more a failure in understanding and interpretation between human and computer. The rapid rise in the use of in-car satellite navigation with its novel dynamic map of the driven world coming into being just beyond

Figure 12.3 Typical newspaper story reporting driving mistakes 'caused' by Satnav mapping errors. Source: author scan from *The Metro*, 2006.

the windscreen is a fascinating illustration of this interpretative failure that has led to a considerable amount of press coverage (Figure 12.3). Map studies might seek to get behind the headlines of these satnav 'cockup' stories to reveal how people cope with this of-the-moment wayfinding mapping combined with turn-by-turn voice instructions. As such, investigating the processes of getting lost may well be more productive than researching successful navigation!

Moments of change and decision making

Where mapping is involved in decision making it does so because it makes a difference. Identifying when maps appear in these processes and assessing the contributions they make is, we would argue, a potentially rich field of research, which might allow researchers to track between representational and non-representational approaches to the world in ways that are 'more-than-representational', linking practices to artefacts and material culture (Lorimer 2005). Monmonier (1996b) offers a useful starting point with its consideration of 'carto-controversies': moments and processes where mapping has been strongly contested.

The role mapping plays in the construction and maintenance of different global world orders, and its contributions to moments of change such as revolutions, boundary disputes or regime change is seriously under-researched. Productive examples illustrating this potential are Crampton's (2006) work on the role of mapping in the inquiry at the end of World War I and Campbell's (1999) consideration of mapping in the Dayton Peace Accord after the Bosnian conflict. The role of maps in navigation and travel is also clearly amenable to this kind of treatment. Here map studies could usefully draw on the experience of mobilities researchers with their focus on the contingent and relational ways in which space is produced through movement (Sheller and Urry 2006). The iconic power of mapping has also been an important force in the progress of intellectual decisions, with visualization at times coming to represent change in intellectual fashion, and at times being strongly influential in changing ways of understanding ideas in many different disciplines. In geography for example, two of the authors are identifying the 'Maps that Matter',[3] charting the ways in which ideas come to be embodied in map form and how this has a lasting impact ion the world of ideas.

The rhythms of mapping

Map studies could also focus on the shape of the patterns of mapping in time–space using the notion of rhythm analysis (developed, in part, by Lefebvre 2004). This theoretical perspective is beginning to pick up traction in human geography, because as Edensor and Holloway (2008) argue '[i]t foregrounds the processual, dynamic and complexity of both space and time, and their imbrication with each other. . . . rhythmanalysis can highlight the experience of both mobility and situatedness, and the ways in which they are blended'. The rhythms of how mapping appears and disappears in everyday activities could be a productive area to research, for example, the meanings of the repeating nightly viewing of the weather map on television, always subtly different, but reassuringly the same. The extent to which mapping always depicts novelty, bringing possible futures into the present and offering alternatives, itself has a temporality, frequency and spatiallity.

Willim (2007: 8) also argues for a more temporally dynamic approach to the analysis of mapping software, noting:

> [t]he uses of these more dynamic technologies transform social and cultural patterns and processes. The software-based map of GPS-devices represent space not only as distances and spatial relations but also as rhythmic patterns. These technologies may combine spatial and temporal representations in new ways which highlights human experience of the spatial as something also temporal.

Memories of the moments of mapping

Mapping has always evoked memories, leaving traces behind of its reading that resonate in the everyday experience of individuals in different societies. Anthropological approaches to mapping argue strongly that these traces play important but understated roles in the construction of identities, in senses of place and in practical wayfinding skills (Ingold 2000). Memories of paper mapping have been captured in narrative (see Harley 1987). The digital transition affords new research possibilities for investigating these traces of past practice. What we see as a stable map interface on our screens is really a provisional instantiation of algorithms and data, fundamentally ephemeral and unstable, made-of-the-moment and disappearing as quickly as electrons are switched and pixels fade. These fleeting map interfaces, that emerge from software spaces, leave new kinds of traces of their presence in the world, a pattern memory of their creation preserved in automatically generated logs of the executing code. These logs can themselves be rendered visually, as maps of map memories revealing when and where people are mapping their worlds. As an example that illustrates, in a rudimentary fashion, the potential of these map memories is the 'heatmap' created by Fisher (2007) showing the differential interest levels of users of Microsoft's Virtual Earth mapping systems (Figure 12.4; see also Aoidh *et al.* 2008). The previously apparently fixed map interface can itself be charted as the memories embedded in its construction are themselves also available: for example, the explosive growth of OpenStreetMap is mapped as an animation, made up of individual mapping stories brought together into a moving set of mobile memories.

The degree to which significant moments of mapping are automatically captured in memories of map use and construction needs to be researched. This empirical work would inevitably have serious ethical implications because of the risks that these memories reveal much more than intended (e.g. searching for the address and directions to an abortion clinic). It also seems likely that the nation state and corporations will be interested in the surveillant potential of individual logs of geographical search and online mapping. The mundane, yet intimate, scope of tracking of social lives from our moments of mapping is part of a wider concern that the world of code does not forget (cf. Dodge and Kitchin 2007).

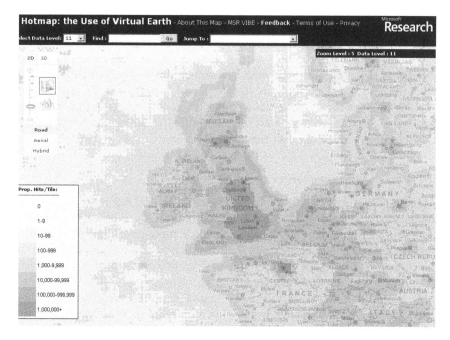

Figure 12.4 Memories of mapping. Source: author screenshot from http://hotmap.
          msresearch.us/.

Mapping ourselves – moments in academic practices

As an introspective moment, map studies could explore how academics, including geographers, deploy maps in their everyday praxis, in university laboratories, their offices and lecture halls. Ongoing questioning of the relation between academic geography and the map could be a productive area to research, leading to a more critical geography of cartography, exploring more than simply publications and curriculae (cf. Dodge and Perkins 2008). It can be argued that there has been disappointingly little development in terms of progressive and creative *use* of maps by human geographers in their researches; Perkins (2004: 385) laments: '[d]espite arguments for a social cartography employing visualizations to destabilize accepted categories most geographers prefer to write theory rather than employ critical visualization'. The humanistic cartography of Danny Dorling is a notable exception to this (the Worldmapper cartogram project he leads has enjoyed considerable success and widespread use). Dorling (2005) has argued for socially informed mapping to educate the next generation of geographers and also to influence public policy by more effectively and creatively highlighting the extent of social inequalities across space; '[m]aps are powerful images', acknowledges Dorling (1998: 287), but this can be exploited in a progressive way, '[f]or people who want to change the way we think about the world, changing our maps is often a necessary first step'. Map studies needs to explore these educative moments of mapping in schools and universities.

Creative moments

A common current in post-structural thought emphasizes that the world may be better theorized as a series of interlinked and constantly changing flows, as a network of possibilities, as a series of bounded possibilities in which change is the only constant and where immanence comes to replace essence (Massey 2005). Map studies needs to create new ways of mapping this context. We live in a time of unprecedented mapping possibilities, in which more people than ever before are engaging in mapping, making their own maps and deploying mapping in novel ways. Artists are deploying the map more than ever before to explore our relationship to the world. Writers use cartographic metaphors to express many different ideas about place. Filmmakers constantly return to mapping as a motif for the human condition. But this mass everyday explosion of mapping is largely taking place outside of the world of map studies. We argue that the creative possibilities of all this new mapping ought to inform our studies too, and that we ought not to separate the analytical from the creative. People studying maps in creative ways need to be more creative in their mapping activities as well.

Conclusion

The world is changing and the way we understand these changes is itself making new worlds. Mapping is part of this process: maps are products of the world and they produce the world. Such changes demand a new manifesto – new ways of thinking, researching and creating maps. For too long, much mapmaking and research has replicated old certainties, focusing on areas, scales and themes, deploying rather tired existing ways of imagining the world and simply applying these to interactive, animated and multimediated contexts, instead of exploring the full potential of new contexts, styles and technologies. As we have argued in this chapter, and as the various chapters in this volume demonstrate, rethinking the modes, methods and moments of maps offers a myriad of new, productive ways to progress cartographic theory and praxis.

  As we have collectively argued and illustrated, alternatives need to be made and worked through that push cartography beyond the pursuit of refining itself as a set of ontic knowledges (where the map has essential qualities that are improved solely through technical advancements; see Chapter 1). Our arguments in this concluding chapter have accordingly highlighted *what* aspects of these changing intellectual landscapes may be particularly worthy of attention, identifying some possible ways forward, flagging up some of the many possible options in *how* the new terrains may be studied, and trying to contextualize this manifesto by stressing that all research needs to be situated, placed and timed. Research and rethinking are both processes, and although in the words of the song, the future's not ours to see, mapping has always been particularly good at bringing it home, offering a route through the infinity of possible outcomes. So to conclude this narrative demands a call for action – a new manifesto: rethink and remake your map studies and practice!

Notes

1   Ritzer (2008) discusses the genealogy of shifts towards a prosumer model of capitalism, in which prosumers produce at least part of what they consume.
2   This denial is, of course, not universal. Researchers in the history of cartography community in particular have long maintained a deep concern with the materiality of cartographic objects. This concern is in terms of both the qualities of the materials used in map production (here primarily as evidence, e.g. for identification of the origins, dating and claims of authenticity; and for the optimal means of preservation and conservation of artefacts themselves) and also the importance of embodied interactions and 'connection' with maps as an innate part of deep interpretative scholarship and the connoisseurship of the collector (the affective feel of holding old maps in particular, the emotional need to be in direct touch with original materials).
3   Some initial ideas are presented as a blog, http://mapsthatmatter.blogspot.com/.

References

Aitken, S. and Craine, J. (2006) 'Affective geovisualizations', *Directions: A Magazine for GIS Professionals*, 7 February, <www.directionsmag.com/printer.php?article_id=2097>.
Anderson, B. and Harrison, P. (2006) 'Commentary: Questioning affect and emotion', *Area*, 38(3): 333–5.
Aoidh, E.M., Bertolotto, M. and Wilson, D.C. (2008) 'Understanding geospatial interests by visualizing map interaction behavior', *Information Visualization*, 7: 275–86.
Batty, M. and Hudson-Smith, A. (2007) 'Imagining the recursive city: Explorations in urban simulacra', in H.J. Miller (ed.) *Societies and Cities in the Age of Instant Access*, Dordrecht, Netherlands: Springer, 39–55.
Beck, R.A. (2003) 'Remote sensing and GIS as counterterrorism tools in the Afghanistan war: A case study of the Zhawar Kili region', *The Professional Geographer*, 55: 170–9.
Bolter, J.D. and Grusin, R. (1999) *Remediation: Understanding New Media*, Cambridge, MA: MIT Press.
Brown, B. and Laurier, E. (2005) 'Maps and journeys: An ethno-methodological approach', *Cartographica*, 40(3): 17–33.
Campbell, D. (1999) 'Apartheid cartography: The political anthropology and spatial effects of international diplomacy in Bosnia', *Political Geography*, 18(4): 395–435.
Chen, B.X. (2008) 'Google's super satellite captures first image', *Wired News*, 8 October, <http://blog.wired.com/wiredscience/2008/10/geoeye-1-super.html>.
Clark, N., Massey, D.B. and Sarre, P. (2008) *Material Geographies: A World in the Making*, London: Sage for the Open University.

Cloud, J. (2002) 'American cartographic transformations during the Cold War', *Cartography and Geographic Information Science*, 29(3): 261–82.

Crampton, J.W. (2003) *The Political Mapping of Cyberspace*, Edinburgh: Edinburgh University Press.

Crampton, J.W. (2004) 'GIS and geographic governance: Reconstructing the choropleth map', *Cartographica*, 39(1): 41–53.

Crampton, J.W. (2006) 'The cartographic calculation of space: Race mapping and the Balkans at the Paris Peace Conference of 1919', *Social and Cultural Geography*, 7(5): 731–52.

Crampton, J.W. (2009) 'Maps 2.0', *Progress in Human Geography*, 33(1): 91–100.

Diakopoulos, N., Luther, K., Medynskiy, Y. and Essa, I. (2007) *The Evolution of Authorship in a Remix Society*, mimeo, <www.deakondesign.com/wp-content/uploads/2007/06/hts5-diakopoulos.pdf>.

Dix, A.G., Finlay, J., Abowd, G.D. and Beale, R. (2004) *Human–Computer Interaction*, 3rd edn, London: Prentice Hall.

Dodge, M. (2008) *Understanding Cyberspace Cartographies*, unpublished PhD thesis, Centre for Advanced Spatial Analysis, University College London.

Dodge, M. and Kitchin, R. (2004) 'Flying through code/space: The real virtuality of air travel', *Environment and Planning A*, 36(2): 195–211.

Dodge, M. and Kitchin, R. (2007) '"Outlines of a world coming in existence": Pervasive computing and the ethics of forgetting', *Environment and Planning B: Planning and Design*, 34(3): 431–45.

Dodge, M. and Perkins, C. (2008) 'Reclaiming the map: British geography and ambivalent cartographic practice', *Environment and Planning A*, 40(6): 1271–6.

Dodge, M. and Kitchin, R. (2009) *Code/Space: Software, Space and Society*, Cambridge, MA: MIT Press.

Dodge, M., McDerby, M. and Turner, M. (2008) *Geographic Visualization: Concepts, Tools and Applications*, London: Wiley.

Dodson, S. (2005) 'Get mapping: As mapmaking becomes big business, citizen cartographers are creating free personal alternatives', *The Guardian*, 7 April, <www.guardian.co.uk/technology/2005/apr/07/onlinesupplement3>.

Dorling, D. (1998) 'Human cartography: when it is good to map', *Environment and Planning A*, 30: 277–88.

Dorling, D. (2005) *Human Geography of the UK*, London: Sage.

Dykes, J.A., MacEachren, A.M. and Kraak, M.-J. (2005) *Exploring Geovisualization*, Amsterdam: Elsevier.

Edensor, T. and Holloway J. (2008) 'Rhythmanalysing the coach tour: the Ring of Kerry, Ireland', *Transactions of the Institute of British Geographers*, 33: 483–501.

Edney, M.H. (1993) 'Cartography without "progress": Reinterpreting the nature and historical development of mapmaking', *Cartographica*, 30(2/3): 54–68.

Elkins, J. (2003) *Visual Studies: A Skeptical Introduction,* New York: Routledge.

Erle, S., Gibson, R. and Walsh, J. (2005) *Mapping Hacks: Tips & Tools for Electronic Cartography*, Sebastopol, CA: O'Reilly & Associates, Inc.

Fisher, D. (2007) 'How we watch the city: Popularity and online maps', *Workshop on Imaging the City, ACM CHI 2007 Conference*, <http://research.microsoft.com/~danyelf>.

Fuller, M. (2008) *Software Studies: A Lexicon*, Cambridge, MA: MIT Press.

Geller, T. (2007) 'Imagining the world: The state of online mapping', *IEEE Computer Graphics and Applications*, March/April, 8–13.

Gill, S. (2004) *Field Studies*, London: Chris Boot.

Graham, S. (2005) 'Software-sorted geographies', *Progress in Human Geography*, 29(5): 562–80.

Graham, S. and Thrift, N. (2007) 'Out of order: Understanding repair and maintenance', *Theory, Culture & Society*, 24(3): 1–25.

Haklay, M. (2008) *How Good is OpenStreetMap Information? A Comparative Study of OpenStreetMap and Ordnance Survey Datasets for London and the Rest of England*, mimeo, <http://povesham.wordpress.com/2008/08/19/openstreetmap-quality-evalution-and-other-comparisons>.

Harley, J.B. (1987) 'The map as biography: Thoughts on Ordnance Survey map, six-inch sheet Devonshire CIX, SE, Newton Abbott', *The Map Collector*, 41: 18–20.

Harley, J.B. (1988) 'Maps, knowledge and power', in D. Cosgrove and S. Daniels (eds) *The Iconography of Landscape*, Cambridge: Cambridge University Press.

Harley, J.B. (1989) 'Deconstructing the map', *Cartographica*, 26(2): 1–20.

Harrower, M.A. and Brewer, C.A. (2003) 'ColorBrewer.org: An online tool for selecting color schemes for maps', *The Cartographic Journal*, 40(1): 27–37.

Harvey, D. (1989) *The Condition of Postmodernity*, Oxford: Blackwell.

Ingold, T. (2000) *The Perception of the Environment: Essays on Livelihood, Dwelling and Skill*, London: Routledge.

Introna, L.D. and Ilharco, F.M. (2006) 'On the meaning of screens: Towards a phenomenological account of screenness', *Human Studies*, 29: 57–76.

Kaplan, C. (2006) 'Precision targets: GPS and the militarization of U.S. consumer identity', *American Quarterly*, 58(3): 693–714.

Kitchin, R. and Dodge, M. (2007) 'Rethinking maps', *Progress in Human Geography*, 31(3): 331–44.

Kurgan, L. (1994) 'You are here: information drift', *Assemblage*, 25: 14–43.

Kwan, M.P. (2007) 'Affecting geospatial technologies: Toward a feminist politics of emotion', *The Professional Geographer*, 59(1): 22–34.

Lee, K.D. and Shumakov, A. (2003) 'Access to geospatial data in 2003: a global survey of public policy and technological factors', *Cartography and Geographic Information Science*, 30(2): 225–30.

Lefebvre, H. (2004) *Rhythmanalysis: Space, Time and Everyday Life*, London: Continuum.

Longan, M.W. (2008) 'Playing with landscape: Social process and spatial form in videogames', *Aether*, 2: 23–40.

Lorimer, H. (2005) 'Cultural geography: The busyness of being more-than-representational', *Progress in Human Geography*, 29(1): 83–94.

Lyon, D. (2007) *Surveillance Studies: An Overview*, Cambridge, Polity Press.

Manovich, L. (2005) 'Remix and remixability', *Nettime*, 16 November, <www.nettime.org/Lists-Archives/nettime-l-0511/msg00060.html>.

Massey, D.B. (2005) *For Space*, London: Sage.

Monmonier, M. (1996a) *How to Lie with Maps*, 2nd edn, Chicago: University of Chicago Press.

Monmonier, M. (1996b) *Drawing the Line: Tales of Maps and Cartocontroversy*, New York: Henry Holt.

Norman, D. (1998) *The Invisible Computer*, Cambridge, MA: MIT Press.

O'Loughlin, J. (2005) 'The war on terrorism, academic publication norms, and replication', *The Professional Geographer*, 57(4): 588–91.

Parks, L. (2004) 'Kinetic screens: Epistemologies of movement at the interface', in N. Couldry and A. McCarthy (eds) *Mediaspace: Place, Scale and Culture in a Mobile Age*, London: Routledge, 37–57.

Pearce, M.W. (2008) 'Framing the days: Place and narrative in cartography', *Cartography and Geographic Information Science*, 35(1): 17–32.

Perkins, C. (2004) 'Cartography – cultures of mapping: power in practice', *Progress in Human Geography*, 28(3): 381–91.

Perkins, C. (2008) 'Cultures of map use', *The Cartographic Journal*, 45(2): 150–8.

Pickles, J. (1991) 'Geography, GIS, and the surveillant society', *Papers and Proceedings of Applied Geography Conferences*, 14: 80–91.

Pickles, J. (2004) *A History of Spaces: Cartographic Reason, Mapping and the Geo-Coded World*, London: Routledge.

Ritzer, G. (2008) *Production, Consumption . . . Prosumption?*, mimeo, <www.georgeritzer.com/docs/>.

Ronson, J. (2004) 'Attention to detail', *The Guardian*, Saturday Magazine, 15 May, <www.guardian.co.uk/print/0,,4923064–103425,00.html>.

Rowell, J. and Ungar, S. (2003) 'The world of touch: Results of an international survey of tactile maps and symbols', *The Cartographic Journal*, 40(3): 259–63.

Selfe, C.L. and Selfe, R.J. (1994) 'The politics of the interface: Power and its exercise in electronic contact zones', *College Composition and Communication*, 45(4): 480–504.

Sheller, M. and Urry, J. (2006) 'The new mobilities paradigm', *Environment and Planning A*, 38(2): 207–26.

Slavick, E.O. (2007) *Bomb After Bomb: A Violent Cartography*, New York: Charta.

Star, S.L. (1999) 'The ethnography of infrastructure', *American Behavioral Scientist*, 43(3): 377–91.

St Martin, K. (1995) 'Changing borders, changing cartography: Possibilities for intervention in the new world order', in A. Callari, S. Cullenberg and C. Biewener (eds) *Marxism in the Postmodern Age*, New York: Guilford Press, pp. 459–68.

Sturken, M. and Cartwright, L. *(2007) Practices of Looking: An Introduction to Visual Culture,* 2nd edn, Oxford: Oxford University Press.

Taylor, D.R.F. (2005) *Cybercartography: Theory and Practice*, Amsterdam: Elsevier.

Thrift, N. and French, S. (2002) 'The automatic production of space', *Transactions of the Institute of British Geographers*, 27: 309–35.

van Elzakker, C., Nivala, A-M., Pucher, A. and Forrest, D. (2008) 'Caring for the users', *The Cartographic Journal*, 45(2): 84–6.

Willim, R. (2007) *Walking Through the Screen: Digital Media on the Go*, mimeo, <www.robertwillim.com/rw_gps.pdf>.

Wood, D. (2003) 'Cartography is dead (thank god!)', *Cartographic Perspectives*, 45: 4–7.

Zook, M.A. and Graham, M. (2007) 'Mapping digiplace: Geocoded internet data and the representation of place', *Environment and Planning B: Planning and Design*, 34: 466–82.

Index

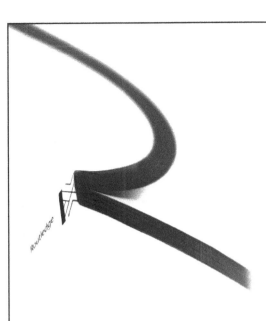